Cosmic Pu

From Atoms to Consciousness

Cosmic Puberty

From Atoms to Consciousness

Neil English

The Lutterworth Press

Cambridge

Dedication

This book is dedicated to the memory of Carl Edward Sagan (1934-1996), a scientist who taught me more about religion than any church preacher could. May he find everlasting peace in the great cosmos to which he has returned.

First published in 1999 by

The Lutterworth Press
P O Box 60
Cambridge
CB1 2NT
England

e-mail: **publishing@lutterworth.com**
internet: **http://www.lutterworth.com**

British Library Cataloguing in Publication Data:
A catalogue record is available from the British Library

ISBN 0 7188 2984 0

Printed in Great Britain by
St Edmundsbury Press Ltd, Bury St Edmunds, Suffolk

Contents

Acknowledgements

I am deeply indebted to all those who read earlier versions of the manuscript, and who provided sound constructive criticisms on its content, especially Dr. Patrick Moore, Dr Stuart Clark, Jeremy Cresswell, John McNicol, Professor Roland Wolf, Margaret Cameron, and Claire Cargill, Michelle Golder and John Woodruff. Grateful thanks are extended to my brother, Michael, for providing some unique suggestions on the temperament of the book. I would also like to thank my wife to be, Lorna Rankin, for keeping me on the straight and narrow these past few months! And to little Leah-Zoe, the source of all my inspiration. Special thanks is extended to Adrian Brink and colleagues at the Lutterworth Press, for their sound advice in the layout and promotion of every aspect of the book.

Preface

I have learned
To look on nature, not as in the hour
Of thoughtless youth; but hearing often-times
The still, sad music of humanity

William Wordsworth (1770-1850)

As far back as I can remember, I have always had an interest in science. My first love was archaeology. As an eight year old boy, I became captivated by the methods archaeologists use to unearth fragments of humanity's past. I was fortunate enough to have been raised in a family that cherished the pursuit of knowledge for its own sake. I quickly absorbed the relevant sections from popular encyclopaedias provided by my parents. If I were lucky, my father would call my attention to interesting *Reader's Digest* articles on the origin of humanity, the Big Bang, the war against cancer and much, much more. From the outset, it was clear that I was hooked on science. For the first time in my life I felt that I was just one small part of something far greater than myself.

At eleven, I became interested in astronomy. I painstakingly memorised the positions and names of the stars making up the brighter constellations visible from my urban front garden in Ireland. After I had constructed a pathetic contraption bearing an uncanny resemblance to a telescope, my sister and brother-in-law took pity on me. They must have sensed that I had an overwhelming desire to peer a little deeper into the mysteries of the universe. Things turned out fine. They presented me with a small, 2-inch diameter telescope, which I optimistically named 'Galileo'.

I was euphoric. I stayed up most of that night to see if the stars or a bright planet would emerge from the thick blanket of cloud. It was not to be. I had grown accustomed to the cloudy Irish climate. In fact, as I recall, I did not have the opportunity to turn my beautiful little 'scope to the heavens for another week. What I saw then changed my life for ever. I decided there and then to become a scientist, to seek to understand the intricate tapestry of truth that lay hidden behind nature's great veil.

As a student, I learned about the universality and elegance of the laws of nature that govern tiny subatomic particles and enormous supergiant stars with equal facility. I hungrily studied the intricacies of biological systems with its finely interwoven biochemical pathways breathing life into inanimate matter. I began to wonder about the origins of life. Whence came the splendour that is the Earth's biosphere, replete with millions of species of plants, animals and microbes? And what of other worlds? I was electrified to discover that the molecules that impart vitality to matter are strewn across the far reaches of the universe, everywhere we look. I was impelled to imagine exotic forms of life, utterly bizarre and far removed from terrestrial biology.

Lost in the immensity of space and time set before us, we humans struggle for an identity. From a very early age, we yearn to be a part of something – a band, a choir, a football team – and as adults we join societies, associations and clubs. We are immensely proud of our cultural identities and our nationality. But our explorations of the universe have made us sit up and take note of our ultimate identity. We are citizens of a vast, ancient and, we have every reason to believe, purposeful universe. We have the immense good fortune to be alive and conscious at the end of the twentieth century. As we stand on the shoulders of towering intellects who paved a path through a jungle of ignorance, we recognise their timeless contributions. But we have not reached the end of our journey. Vast tracks of cosmic wilderness remain unexplored, patiently awaiting our discovery and introspection. *Quo vadimus?* Where are we going?

Deep inside every one of us there is a great, passionate longing to merge with the cosmos. In the still of the night, we watch the stars. We marvel at their fragile beauty. We feel a sense of loneliness for them. The stars hold a timeless elegance and grace made manifest by the contemplations of the human mind. We owe our entire existence to the refulgent stars, objects of great simplicity.

For 12 billion years we have been unconscious travellers, drifting slowly across a great ocean of space. But a million years ago, something wonderful happened. The cosmos stirred. An extraordinary series of events led to the emergence of the genus *Homo*, ancestors to ourselves. The universe became conscious. And slowly, through the methods of science, we have come to understand much about its make-up. I want to share with you my joy of knowing a little about the cosmos. It is a story that starts with atoms and photons of light, and ends with their logical product, consciousness and ourselves.

This is an epoch that has made us aware of our ancient genetic heritage. We already understand the forces of nature that shaped our origin. Does this mean that we have become divorced from these evolutionary forces that have so subtly transformed us into the custodians of our world? We are the products of cosmic evolution, of stellar alchemy. Humanity is one conscious voice that may, as yet, be unheard in the great void between the stars. Living on what may well be an ordinary planet, circling a run-of-the-mill star on the outskirts of a giant pinwheel of stars – the Milky Way Galaxy – we have grown towards cosmic puberty, our inheritance set before us. We have a right of passage to the cosmos and all the wondrous variety it accommodates.

This book starts with a historical description of the scientific ideas that have shaped our modern understanding of the universe. As we turn the pages, we shall venture on a journey to the distant galaxies – the star cities of the cosmos. We cruise among the stars of the Milky Way, exploring the origin of the Sun and its retinue of nine or so worlds. Continuing onwards, we examine the planet Mars as an abode of life, past and present. We voyage to the Jupiter system and take a look at another water world, Europa, hidden beneath a thin crust of ice. The stage is then set to appraise the arguments for and against the existence of extra-terrestrial intelligence, as well as the methods that we use to attempt communication with such beings.

Science has also provided valuable clues to the origins of humankind. We shall unearth tiny, fragmented pieces from the geological jigsaw puzzle to glean some information about our evolution over the past seven million years. As well as venturing out into the macrocosm, the seeming eternity of the 'very big' that confronts us, we also visit the equally inviting microcosm, examining the origin of replicating molecules, the emergence of the first cells and the glorious radiation of biological systems since life's inception, four billion years ago.

Turning from humanity's origins to our future, the book looks at man's life span. This generation has been the first to be enriched by astounding advances in medical knowledge heralded by the revolution in the science of molecular biology. We look at the way people have begun to question age-old beliefs about our apparent mortality, with research demonstrating that we can outlive our biological use-by date. We discover the way in which the very oxygen in the air we breathe conspires with our genes to ordain our demise. As we shall see, oxygen is a Jekyll-and-Hyde molecule, capable of imparting vitality to living cells, but also capable of causing irrevocable damage to the molecular machines at the heart of life.

In a vast and ancient universe our affairs seem petty and insignificant. And in comparison to the legacy of stars that burn ferociously for millions or billions of years, our lives seem to flicker on and off like fireflies in the night. We are finite creatures who, through the collective enterprise of scientific discovery, have unlocked new doors to the past and bridges to the future. Science has brought us face to face with our frailty and our immortality. But with science many of our dreams have come true. We have come to sail the water oceans and the breezes of heaven. We have even dared to leave our planetary home, extending our visions to the Moon and the other planets in our solar system. But our sights are set on an even grander dream – to set sail for the stars, to explore exotic, alien worlds – worlds of ice and rock, desolate worlds too hot or too cold for life, and quite possibly a menagerie of habitable worlds brimming over with life and intelligence. We seek companions across the sea of space.

Yet in coming to grips with the majesty of the universe, in grappling with its immense beauty and age, we sometimes fail to recognise the instrument of all these imposing contemplations – the human mind. By virtue of our consciousness we extract patterns from the havoc of nature and meaning from indifference. I shall argue that the quest to know the universe and whether it contains other thinking creatures, as well as the search for immortality through science, are poignant examples of our most deeply laid desires to connect with that which is beyond us. And although we have come of age on a small, rocky world orbiting an inconspicuous star on the outskirts of a great spiral galaxy, we may rejoice in knowing the universe, in feeling humility and reverence for its illustrious splendour. We are precious beyond imagination.

Puberty marks the transcendence from precocious youth to the maturity of adulthood. Our civilisation is now engaged in this spectacular metamorphosis. Through our astounding scientific achievements we have changed our world irrevocably. I believe that we have reached a critical stage in our cultural and intellectual development. Both are so finely interwoven that it is scarcely possible to disentangle one from the other. We have arrived at our cosmic puberty. What we enact in these future years will, I believe, have powerful repercussions for all future generations. We have a choice. Do we consign our descendants to inhabiting a polluted, anarchistic world, where intellectual darkness reigns supreme, or will we mend our ways and set sail for the stars? In many ways, the future is here and now: too close for comfort!

Chapter I

Our Cosmic Perspectives

The true, strong and sound mind is the mind that can embrace equally great things and small.

Samuel Johnson (1709-84)

Art thou pale for weariness
Of climbing heaven and gazing on the earth,
Wandering companionless
Among the stars that have a different birth,
And ever changing, like a joyless eye
That finds no object worth its constancy?

Percy Bysshe Shelley (1792-1822)

Our existence on this blue planet is nothing short of a miracle. Newton knew it, Einstein knew it. Even Darwin knew it. The laws of nature and the magical properties of atoms have sculpted a cosmos teeming with galaxies, stars, planets and, we have every reason to believe, a glorious riot of living beasts and vegetation. For now, the Earth is unique in our solar system in being the only world that can support such a wondrous variety of life. But we are beginning to see things anew. In our intellectual adolescence we have stumbled upon some of the most basic laws of nature that have overseen the development of life and consciousness on this small world.

The astounding advances in our technological power have allowed us to see deep into the mysterious heart of our origins. Much of what we know about ourselves has been gleaned by studying the greater universe. This chapter describes the nature of the vast and ancient ocean of space that we have the immense good fortune to inhabit. We shall take a look at the physical and chemical forces that led to the formation of the Sun and her attending planets, as well as the evolution of biological organisms.

Human beings have been civilised, at least in part, for 7,000 years. For 94% of this time, we had only the use of our eyes and imaginations to ponder the nature of the heavens and all its resplendent glories. For millions of dark, clear nights our ancestors watched the ice-cold glimmer of the stars. They beckoned to us just as strongly then as they do now. But what could they mean?

An understanding of these marvellous creations of nature grew from the human activity of scientific investigation. By learning about the nature of light, we have unlocked a great many cosmic mysteries. From the moment of our birth, photons of light have powerfully shaped the human perception of the universe. Light is nature's great information super-highway. It is therefore fitting to begin our journey through the universe by considering the alluring properties of light.

Waves of Understanding

Light behaves very strangely. Sometimes it acts as if it were composed of a number of tiny particles and at other times it behaves as if it were a wave. Let's think about the wave theory. Imagine snap-freezing the image of a water wave on the ocean. It has a peak and a trough, the distance between which is called the *pitch* or *amplitude* of the wave. The distance from the peak of one to the next incoming wave is called the *wavelength*. The number of waves that pass a fixed point in a designated period of time is known as the *frequency*. Frequency is related to the amount of energy the wave contains: the greater the frequency of the waveform the more energy it carries. Waves are commonplace in nature. The propagation of sound, water disturbances and radio are all mediated through waves.

Galileo Galilei (1564-1642) first calculated the speed of sound by measur-ing the time the reverberations of a gunshot took to travel between two distant hilltops after sighting the initial spark from the firing of the gun. He presumed (correctly) that light travelled far more swiftly than sound, and that the sighting of the gun spark was essentially instantaneous. Many of Gali-leo's contemporaries thought that the speed of light was incalculable simply because light was presumed to be instantaneous. Indeed, in our everyday experience, light appears to be instantaneous. But appearances can deceive.

In 1675, the Danish astronomer Öle Romer (1644-1710), a keen student of Jupiter and its four large moons, recorded a later-than-expected eclipse of a Jovian moon. At the time, Earth and Jupiter were in conjunction on opposite sides of the Sun and were at their most distant from each other. Romer determined that it took 16 minutes longer for light to travel from Jupiter when Earth was farthest away from Jupiter than when Earth was closest. In a spark of genius, he realised that this 16-minute discrepancy could only be due to the time it took light to cover the distance between the closest and farthest positions of Jupiter. Romer cautiously reported a speed of 240,000 kilometres per second, smaller than the modern figure of 300,000 kilometres per second, but nonetheless colossal enough to enthral many of his scientific contemporaries. The velocity of light, for centuries taken as instantaneous, was now well within the realms of human understanding. Light speed was finite! The significance of this result was incalculable. The enormous velocity of light was to be the subject of powerful ruminations by future generations of scientists, above all Einstein, in the formulation of his epochal theory of relativity.

The invention of the telescope in the first decade of the seventeenth century heralded a new era in visual astronomy, allowing more precise measurement of the skies and extending the horizon of visible objects by orders of magnitude. The primary function of the telescope is to gather light from a distant object and bring the information carried therein to a focus. From the first moment that Galileo turned his primitive spyglass to the heavens in the winter of 1609, the sky was transformed. For the best part of 1,500 years the heavens were held to be changeless. The Moon, Sun and stars were perfections of nature placed in the vault of heaven and destined to follow their paths, unperturbed by petty human affairs. The sky represented a lasting permanence, consonant with an ordered and harmonious cosmos – the antipode of the ephemeral and imperfect world of humans.

The Emancipation of Science

What Galileo saw and faithfully recorded through his telescopes shattered this view, utterly and for ever. Instead of seeing a perfectly smooth Moon, his mortal eyes witnessed a severely cratered and mountainous lunar vista. The Moon, it turned out, was as imperfect as the Earth. His magnificent drawings of the phases of the waxing and waning Moon continue to capture the imagination of budding astronomers. Galileo was arguably the first human being to sight the four bright Jovian satellites. Their ever-changing positions night after night enthralled him. Here, before his very eyes, the distant planet Jupiter with its retinue of lesser worlds confirmed the heliocentric theory of the Polish astronomer Nicholas (1473-1543), who had formulated a theory in which the five known planets and the Earth followed circular orbits about the Sun. This deeply significant discovery was also to provide observational evidence for a theory of gravity later to be formulated by Newton.

Although Galileo made gallant attempts to formulate physical laws to describe the motion of earthly bodies he never applied them to the astronomical discoveries that he made in that fateful year of 1610. In a series of newsletters, distributed throughout Europe and called *Starry Messenger*, Galileo went on to describe the miniature phases of Venus, as well as the appearance of the star-studded veil of the Milky Way. For millennia, mortal human eyes saw only a smudge of light extending across the night sky. But Galileo's telescope transformed the Milky Way into an enormous swarm of stars. Through his mighty little telescope, the wondrous variety and glory of the cosmos was extended beyond imagination. The science of astronomy has never since experienced a time of such fruitful discovery, graced with unbridled, scientific revelation.

The next great step to understanding the nature of the universe was taken by the German astronomer Johannes Kepler (1574-1630). Kepler immersed himself in a quest to decipher the laws governing the motion of planets orbiting the Sun. In particular, he sought to explain the to-and-fro motion of the planet Mars in its path across the sky. To do so he required the best available observational data of the planet's location night after night as it orbited the Sun. He was led to the finest observational astronomer in the known world – the man with the golden nose, Tycho Brahe (1546-1601). In his youth, Tycho's bumptious self-confidence led him to lose his nose in a duel with an adversary over an argument about who was the greater mathematician.

Tycho's observations had led him to conclude that Mars went around the Sun in a perfectly circular orbit. Except for a tiny discrepancy of 8 seconds of arc – an angle less than a quarter of the apparent diameter of the Sun as seen from the Earth's surface – the orbit of Mars could easily have been reconciled with a circular orbit. But this was too large for Kepler to ignore because it was larger than the widest margin of error calculated by Tycho. In courageous decisions, Kepler decided to attempt to fit the orbital data of Mars to different geometrical shapes based on the other conic sections – the ellipse, the parabola and the hyperbola. Finally, in 1609, after years of desperation and endless mathematical calculation, he confirmed that the notion of a circular orbit was in error when he correctly fitted the data to an elliptical circuit. These discoveries formed the basis of his famous three laws of planetary motion. Here they are:

Law 1: Planets trace out an elliptical orbit with the Sun at one focus.

Law 2: Planets sweep out equal areas in equal times in their orbit about the Sun.

Law 3: The square of the planet's orbital period – the time taken to complete one orbit of the Sun – is proportional to the cube of its average distance from the Sun.

Kepler's discovery of the elliptical planetary orbit caused a revolution in science and philosophy. Conventional wisdom held that the celestial bodies could move only in perfectly circular orbits. For millennia, the circle had been revered as a metaphor for the divine. Its beautiful, perfectly symmetrical form was wholly consonant with the changeless vault of heaven. But although Kepler's laws fitted the data for all the planetary orbits beautifully, he was unable to provide any theoretical explanation for these relationships. It was left to the mind of another great thinker to provide the answer. His name was Isaac Newton.

A Revolutionary Theory

The life of Isaac Newton (1642-1727) was long and fruitful. A neurotic, suspicious and tormented misanthrope by nature, Newton spent most of his life in contemplative isolation. His great ruminations during his time as an undergraduate at Trinity College, Cambridge would change the world irrevocably. In his first year at university, Newton invented the differential and integral calculus (actually he is recognised as co-discoverer with the German mathematician and philosopher, Gottfried Wilhelm von Leibnitz (1646-1716), a hugely important branch of mathematics dealing with rates of change, areas and sums. He also formulated the laws of gravitation and motion, which he expounded in his earth-shattering book *Philosophiae Naturalis Principia Mathematica (The Mathematical Principles of Natural Philosophy)*, first published in 1687. He was also a devout Christian and a renowned biblical scholar. Indeed, it is claimed that he owned 31 editions of the Bible in several different contemporary and ancient languages and devoted most of his intellect to theological and alchemical preoccupations. It is calculated that Newton penned over 2 million words on these subjects alone – far more than he wrote on the sciences.

Using the differential calculus, or as he referred to it, 'fluxions', Newton formally deduced Kepler's laws, thus providing an elegant theoretical underpinning for the dynamics of planetary orbits. Newton realised that the gravitational force attracting two bodies together was directly proportional to their masses and inversely proportional to the square of the distance separating them. This makes perfect sense. The larger the mass of an object, the more attractive power it should possess, and the farther apart two objects are, the more attenuated the force of attraction between them should be. Such an 'inverse square law' was found to hold true for many other relationships, such as the force between two point charges in Coulomb's law, and how the intensity of light falls off with distance. It was clear that Newtonian physics represented a profound advance in our knowledge of the cosmos. Indeed, to a large extent, Newtonian physics is responsible for the very existence of our technological society, even if his theory is not true.

The Abdication of Newtonian Mechanics

Newton's physics inspired countless writers, poets and scientists in the Age of Enlightenment. His ideas prevailed for the best part of 250 years. But Newtonianism was eventually superseded at the beginning of the twentieth century by the ruminations of another intellectual giant – Albert Einstein (1879-1955). Widely regarded as one of the greatest scientific geniuses of all time, Einstein revolutionised our understanding of the universe. Newton's laws firmly established the idea of *absolute* space and time. By 'absolute', Newton meant that space and time exist independently of all material objects, including human existence. Remove all the stars and planets from the vault of heaven, Newton argued, and a three-dimensional void would still exist.

Einstein was deeply moved by the concept of *relative motion*. You may be reading this book in the leisurely confines of your own home. You appear motionless, but the Earth is spinning on its axis at a speed of 1670 kilometres per hour. In turn, the Earth is orbiting the Sun at the enormous speed of 110,000 kilometres per hour. If this is not impressive enough, our Sun circles the galactic centre at an astounding 800,000 kilometres per hour, our Galaxy is drifting among the other galaxies, and so on. We have always been unconscious travellers. How much further do we have go before we can say for sure how fast we are travelling in space? What is our *absolute* speed? Einstein wondered if there was a position, privileged over any other that would allow an observer to measure *absolute motion*. These were central ideas in his epochal work in special relativity.

In 1905, Einstein published a paper entitled, 'On the Electrodynamics of Moving Bodies', and its predictions were literally mind-boggling! Einstein had to reject the concept of absolute space and time. Even the everyday quantities of mass and length were not fixed, not absolute in any way, but had values that depended upon the speed at which the individual was travelling. At ordinary speeds, an object displays no appreciable changes in mass or length, but at faster and faster speeds, close to the speed of light, strange effects begin to manifest themselves to a stationary observer. The apparent mass of an object increases rapidly as light speed is approached. Conversely, the observed length of an object would contract. For example, it can be calculated that a 1kg bag of sugar travelling at 98% the speed of light would have an apparent mass of 5kg, while at the same speed, a 30cm ruler would measure only 6cm long! The closer we approach light speed the more pronounced these changes become.

But what about time? How is our perception of time altered at speeds approaching that of light? Einstein showed that, like mass, time is stretched out or *dilated*. Clocks in motion tick out time more slowly than stationary clocks. For example, 1 second on a stationary clock would be stretched out to 5 seconds on a clock travelling at 98% of the speed of light. The body of knowledge that describes all these effects is called the *special theory of relativity*. Relativistic time dilation has been the subject of much debate by physicists over the last century. The most famous and far-reaching relativistic effect is that described by the *twin paradox*, which tells how two young, identical twins are separated; one remains on Earth, while the other boards a starship capable of travelling near the speed of light. Many years later the twins re-unite. But something strange has indeed occurred.

The twin who was left behind to live life under the auspices of *proper time* – the passage of time as you and I experience it – has aged substantially more than the twin who set out on a relativistic voyage to the stars. Now old and frail, she greets her sister who is full of the joys of youth. In conversation, they exchange their respective experiences of the passage of time. They both admit that they experienced nothing out of the ordinary. Nonetheless, the effects of relativity are all too apparent. It will be many years before they will become fully re-acquainted with each other. By travelling fast into space, we alter our perception of time.

Gravity's Not What It Used to Be!

In 1915, Einstein advanced his *general theory of relativity*, which focused on accelerated motion, yielding utterly new insights into the phenomenon of gravity. In a revolutionary insight, Einstein taught us to view gravitation as a property of space, rather than a physical force in the traditional sense. As a result of the presence of matter, space becomes curved. The more massive the object, the more arched space becomes. Using this analogy, a *black hole* would be a region of space that resembles a bottomless pit. Einstein's elegant theories are all well and good, but how do we really know that they are in some sense true? And if they are, what is wrong with Newtonianism?

Surprisingly, the validity of Einstein's theories was widely accepted in an unusually short time. This was undoubtedly due to his presence of mind in offering predictions for both his theories. First, in regard to the special theory of relativity, Einstein proposed that radioactive atoms travelling near the speed of light would decay more slowly. Also, he predicted that the mass of a subatomic particle would increase as it was accelerated towards light speed. Experiments performed in particle accelerators since the 1950s have shown beyond all reasonable doubt that both relativistic time dilation and mass increase do indeed occur at speeds approaching that of light.

The evidence supporting Einstein's general theory of relativity came a lot earlier. Less than four years after its publication in 1915, a total eclipse of the Sun unveiled the validity of his ruminations. Einstein predicted that light from distant stars would curve slightly as it passed near the Sun. This suggested a relatively easy experiment. In 1919, The British Royal Astronomical Society, under the auspices of the pre-eminent English astronomer Arthur Eddington (1882-1944), organised an expedition to a small Portuguese-owned island off the coast of western Africa. They came to witness a Total Solar Eclipse which would allow them to measure the positions of stars located at or around the unmasked solar corona (the outermost layers of the Sun's atmosphere). These observations clearly showed that the stars' position had shifted. The Sun's gravity is powerful enough to cause a *dip* in the curvature of space, allowing scientists to observe the bending of star light. Since then many more experiments, sometimes arising from very unusual circumstances, have helped to consolidate relativity theory as the most precise approximation to the nature of space and time.

Einstein's radical new theory of gravity has also been spectacularly demonstrated through the phenomenon of *gravitational lensing*. Because matter can warp the surrounding space, it bends and lengthens the path of light rays originating either behind or in front of it. It follows that the bigger the mass, the greater the bending. So matter can act as a lens, focusing more distant light rays into a finer and more concentrated image. Many astronomers initially expressed surprise to discover 'pairs' of identical galaxies in many deep sky images. But the frequency with which these optical pairs appeared was far too high to be a chance occurrence. In reality, what astronomers were seeing was the effect of gravitational lensing, allowing two or even multiple images to be produced.

Gravitational lensing has proved enormously useful to cosmologists and astronomers alike. This is because very distant objects, such as the enormously distant quasars – galaxies with gigantic black holes at their centres, pouring tremendous quantities of energy into the void of inter-galactic space – are so faint they are very difficult to spot. But if a massive galaxy located along our line of sight happens to occupy the intermediate space between us and the quasar, the intervening galaxy concentrates and brightens the image, allowing these ancient structures to be studied with greater facility. Nature thus gives us the ultimate tele-scope, a device forged from matter's interaction with the laws of nature, allowing humans to peer back billions of years in time.

Yet another prediction of general relativity is that clocks will run slower if moved into a stronger gravitational field. Suppose we carried a clock from the surface of the Earth to a great height. Because the Earth's gravity diminishes with altitude, our clock would tick out the passage of time more swiftly than on terra firma. This effect was spectacularly demonstrated in 1976, when a rocket carried a clock to an altitude of nearly 10,000 kilometres. The clock was found to gain nearly a billionth of a second every second over the same clock ticking out time at ground level.

It's true, Newtonian mechanics has humbly taken a back seat to rel-ativity theory. In fact, physicists now consider Newton's laws to be good approximations of the greater truth of Einstein's theories. Newtonian mechanics is wholly adequate to allow us to send satellites into Earth-orbit and robotic emissaries to the distant planets of our solar system. On Earth, a world travelling at the abated, 'snail's' pace of 220 kilometres per second around the centre of the great Milky Way Galaxy, we are denied the opportunity, for now at least, to experience relativistic effects. For now, we have to accept the sojourns of the tiny atoms and the great fleeing galaxies as testaments of the validity of Einstein's wonderful postulates.

For me, the most singular conceptual achievement of Einstein's work was the idea that time imparts yet another dimension to the familiar three that we experience. The ancient Greek mathematician Euclid gave us the dimensions of length, breadth and height. We can easily conceptual-ise up and down, from side to side, and backwards-forwards, but how can a fourth dimension be interpreted? To gain some appreciation of what it is like to live in a four-dimensional universe, we need to temporarily put aside our concept of three-dimensional existence.

Suppose instead that we lived in a cosmos with only two dimensions. We can look from side to side all the way round but we can never experience up or down. Imagine that you are stranded in the middle of a great, open, featureless desert. Your gaze stretches toward the horizon in every direction. You might get the impression that you are the centre of something, that somehow, your position is privileged. This is precisely what astronomers see when they observe the galaxies fleeing away from us in all directions. In reality, we are not located at the centre of anything. An observer located in any galaxy in the universe would observe precisely the same celestial recession as us. Now think of our two-dimensional world as inhabiting the surface of an inflating sphere, like a balloon. Our Galaxy, and all the other galaxies in the universe, can be seen to exist in two dimensions on the surface of the balloon. The increasing *radius* of the balloon represents the extra dimension of *Time*, pushing the galaxies out into the future.

Light Year After Light Year

The light-gathering power of modern astronomical telescopes has given us a tool by which we extend our visions of the cosmos by orders of magnitude. Inside the solar system, light takes only a matter of hours to travel from the dusky skies of distant Pluto to the Earth. Think of the Sun for a moment. Imagine that we are travelling at the speed of light away from its fiery surface into the cool of interplanetary space. Let's head in the direction of the Earth. How long would it take you to get home? The answer, it turns out, is about eight minutes. If our star were fated to follow an explosive demise, we would not know of it for eight minutes, at which time the end of the world would surely be nigh.

The stars are much further away. The nearest ones are located in a cluster in the constellation of Centaurus in the southern hemisphere. The soft yellow and white glimmer from these distant stars takes over four years to reach us. We say that these stars are over four *light years* away. This is an immense distance – over 40 trillion kilometres (a trillion is a 1 followed by 12 zeros). To get some measure of appreciation of these achingly vast distances, we can look to the fastest vehicles ever designed by the human species; the Voyager 1 and 2 spacecraft. These superbly constructed robotic emissaries, which revolutionised our knowledge of the outer solar system, are currently travelling at a velocity of 80,000 kilometres per hour out of the solar system towards the distant stars. They may soon reach the edge of the Sun's dominion, where the gentle breezes of energetic, charged particles, the *solar wind*, meet with those of the nearby stars. This region of interplanetary space is called the *heliopause*. At that point, even if the Voyager spacecraft were headed towards the 'nearby' Centaurus system, they would take many tens of thousands of years to reach it. Perhaps, by then, our memory of them will have grown dim in the archives of human achievement. Perhaps we will consider the significance of the launch of these semi-intelligent machines as no more than an academic curiosity, rather like our contemporary consideration of the early Neolithic people who fashioned primitive stone tools.

Most stars that we can see in our night sky are only a few tens to a few hundred light years distant. The great luminous ball of gas shining below the belt in the constellation of Orion, the magnificent Orion Nebula, is located 1,500 light years away. Further away still is the great galaxy in Andromeda, over 2 million light years distant. Andromeda is a giant galaxy like our own Milky Way, and a member of a small group of about 30 star cities including the Large and Small Magellanic clouds, collectively called the Local Group, bound to each other by gravity. When you next look at the fuzzy haze of the great Andromeda Galaxy high in the east on an autumn night, remember that you are witnessing the end of a two million year journey of its light. When these light rays first set out across intergalactic space, mankind was a far cry from being the guardian of our planet. When this light set out, our hominid ancestors, upright and curious, on the plains of eastern Africa, may have already started to wonder about the nature of those pinpoints in the night sky. Seeing, it seems, is a time machine.

Light – Nature's Great Information Superhighway

Isaac Newton demonstrated that white light is not pristine but is instead made up of a number of colours. For many centuries philosophers had seen the colours of the rainbow but never connected it to white light. So far as we know, Newton was the first to direct sunlight, by means of a curtain slit in his darkened room, through a triangular glass prism to produce the beautiful colour dispersion that he referred to as a *spectrum*. White light was somehow broken down or *resolved* into the colours of the rainbow: red, orange, yellow, green, blue, indigo and violet. Here in his very own room at Trinity College, Cambridge, Newton had his own rainbow. A crock of gold was nowhere to be found; yet, ruminating on the meaning of his colourful experimental displays, Newton discovered something far more valuable. He correctly concluded that white light was a composite made up from purer light of different colours. He also noticed that the prism always bent, or *refracted*, red light most and violet least. Other scientists extended the work of Newton by investigating invisible regions of the spectrum located just beyond the coloured edges.

In 1800, the celebrated astronomer and telescope-maker, William Herschel (1738-1822), found that if he placed a mercury thermometer slightly outside the red region of the spectrum, it heated up. This *infrared* radiation was invisible, yet its physical existence was made manifest by careful experiment and reason, the two most important tools of science. Just a year later, the German physicist Johann Ritter noticed that if silver nitrate (used extensively in photography) was placed just beyond the violet end of the spectrum, it was broken down to silver much more rapidly than if it were placed in blue or violet light. We now recognise this form of light as *ultraviolet*. These observations were but the tip of the iceberg. New realms of the spectrum lay in waiting to be discovered.

The experiments of Ritter and Herschel have shown us that outside the visible region of the spectrum there are invisible but still very real forms of 'light'. And why not suppose that beyond the ultraviolet or the infrared there are still more forms of light? We now know that other forms of *electromagnetic* (*EM*) *radiation* exist – including deadly X-rays and gamma rays lying beyond the ultraviolet, and microwaves and radio waves existing beyond the infrared.

In the early years of the nineteenth century, another great revelation of the information carried by starlight was made known to astronomers. In 1814, a German optician took the first step in the unravelling of the chemistry of the stars. His name was Joseph von Fraunhofer (1787-1826). In the tradition of Newton, he passed sunlight through a narrow slit before allowing it to be refracted by a prism, but he then magnified the resulting spectrum in a small refracting telescope of his own design. Fraunhofer was amazed to find the solar spectrum crossed by numerous sharp lines. He analysed these spectral lines and accurately catalogued the positions of over 700 of them. If Newton had bothered to analyse sunlight under high magnification, as Fraunhofer did, he may have discovered these mysterious lines a whole century earlier. Fraunhofer was totally baffled by their nature but was apparently so preoccupied with the construction of the finest achromatic refracting telescopes in the world that he never put forward any sort of explanation for their existence.

It was not until October 1859 that the German chemists Robert Bunsen (1811-99) and Gustav Kirchhoff (1824-87) worked out a method of identifying chemical elements using *spectroscopy*. They heated many different substances using a gas-fuelled burner, devised by Bunsen, and found that metals and their salts always burn with a characteristic, coloured flame. Compounds of sodium and potassium, for example, always produce the colours yellow and lilac, respectively. Further, they realised that it was possible to identify a chemical element by matching it with a colour found in the spectrum of the substance under investigation. But what were the mysterious dark lines hidden away in the light of the Sun?

Bunsen and Kirchhoff were able to show that hot, dense, gaseous objects emit a *continuous spectrum*, like that seen by Newton, and if this light is passed through a cooler, more rarefied gas, it is soaked up, or absorbed at certain wavelengths. The presence of an element is betrayed by the imprinting of a characteristic set of absorption lines across the spectrum. Moreover, the positions of the absorption lines were characteristic for each chemical element – a universal signature of the atom. This immediately suggested a very important use for the discovery. A new tool, called a *spectroscope*, composed of a prism and a telescope, could be used to decipher the composition of incandescent objects. For the first time in history humans had discovered how to use light to decipher the constitution of the stars.

By the 1890s, men such as William Huggins (1824-1910) and Edward Charles Pickering (1846-1919) were studying many stellar spectra, and on the basis of this analysis, stars began to be classified into spectral types. Originally, stars with particular spectral signatures were classified by letters of the alphabet. This notation has been upheld to the present day. Because most of the visible radiation from stars comes from a relatively thin layer about 500 kilometres thick, astronomers refer to this as the 'surface' of the star. If we arrange stars in order of decreasing 'surface' temperatures then we get this order: O, B, A, F, G, K, M, L. Incidentally, a traditional way to remember this is to assign the mnemonic, "Oh Be A Fine Girl (Guy), Kiss Me Love!" to these letters. Each classification can be further divided into numbers from 1 to 10. For example the Sun has been classified as a G2 star, while one of the nearest stars to our own, Procyon, is considerably hotter, an F5 spectral type.

Knowledge of the spectrum even allowed us to discover new elements, not previously known on Earth. In 1868 the French astronomer Pierre Jules Cesar Janssen (1824-1907) was observing a total eclipse of the Sun over India when he noticed a spectral line that matched nothing seen in any known terrestrial substances. This turned out to be the spectral signature of an inert gas and the second most abundant element in the universe. It was appropriately named *helium*, after the Greek for 'the Sun' (helios) by Norman Lockyer (1836-1920), who independently observed it two months after Janssen. Another thirty years passed before helium was first identified on Earth.

Many physicists in the early twentieth century pondered on the further significance of spectral lines. Perhaps they were a key to the heart of the atom. As we have seen, each element displays a particular set of lines, strung out across the spectrum. For a given substance, this pattern is always the same. The important work of Johann Jakob Balmer (1825-98), who in 1885 suggested that hydrogen produces a whole series of regularly spaced lines, was to prove seminal in the decoding of the atom. Balmer went on to show that the distribution of lines could be neatly described by a simple mathematical rule now described as the *Balmer series*. Complementary work by Lymann and Paschen revealed that there were very similar line patterns in the ultraviolet and infrared regions of the spectrum. They obeyed uncannily similar mathematical rules to that found by Balmer. It was clear that something deeply significant was at work here. Plainly, some great simplicity in atomic behaviour must have underlined this remarkable set of relations.

The work of Ernest Rutherford (1871-1937) and J.J. Thomson (1856-1940) in England did much to establish the physical make up of the atom, which was shown to be composed of oppositely charged particles, the electron and proton. It was not until 1932 that the neutron was discovered by James Chadwick (1891-1974). At the centre of the atom is the nucleus, carrying the positively charged protons and most of the material bulk of the atom. Much further out preside the orbiting, low mass, electrons with their negative electrical charges. These discoveries heralded the birth of atomic physics. It was natural, though arguably naïve, to compare this simplistic notion of the organisation of the atom to that of a miniature solar system. But it was here that physicists, in their reluctance to set aside traditional conventions using classical laws, were met with an insoluble intellectual impasse. They could not reconcile the behaviour of atoms with the laws of Newtonian physics.

Weird Physics

At the end of the nineteenth century, Newtonian physics provided a seemingly cogent, all-encompassing theory of the universe. The solar system was likened to a great clockwork contraption, carefully designed and predictable in every way. Everything we have described in this book so far can be rationalised by these laws, from a falling apple to the motion of stars orbiting the centre of a spiral galaxy. It was natural for a complacency to fall upon physics and physicists alike. But, slowly and imperceptibly, a few contradictions to the formidable doctrine of classical physics emerged. Their bold appearance in the face of assumed cosmic coherency resisted rational explanation by the known laws of physics. Necessity is the mother of invention and the tools to understand the universe were no longer adequate.

As early as the 1880s, some physicists had begun to study the pattern of radiation emitted from hot objects. Everybody knows that if you place an iron poker in an open fire, it will start to glow red, then orange and (if the fire is hot enough) finally yellow. If you had a high-temperature furnace, the iron would eventually glow with a white or even blue-white colour. As we proceed through the visible spectrum from red through blue, there is a shortening of wavelength, the red light having the longest wavelength and blue light the shortest. But as we have seen, the frequencies of light do not end at the blue or red end of the spectrum, but continue on towards the ultraviolet or the infrared, and beyond in both directions.

Astronomers can use this information to deduce the temperature of a star from its colour. This distribution of emitted radiation from an object as a function of wavelength is referred to as a *black-body spectrum*. A black body is an idealised object in physics that acts as a perfect absorber and emitter of radiation. The inside of hot domestic ovens and stars are good examples of near-perfect black bodies. Many attempts were made to explain the shape of their spectra in terms of the known laws of physics. The most formidable attempts to reconcile the behaviour of black-body radiation with classical mechanics were made by Wilhelm Wien (1864-1928) and Lord Rayleigh (1842-1919). Although Wien's scheme worked well for short wavelengths, it failed over longer wavelengths. Rayleigh's theory had the opposite effect. Neither could adequately explain the data obtained from an experimental black body. It was not until 1900 that someone stumbled on an encouraging clue. His name was Max Planck (1858-1948).

Planck proposed that, in order to explain the way in which the wavelength of radiation emitted by a hot black body shifted with temperature, one had to abandon the classical view that the emission of radiation from a hot object was continuous from one temperature to another. Instead, he argued, it was distributed in minute and discrete bundles of energy called *quanta*. Surprisingly, this fitted the observed evidence perfectly, forcing Planck to abandon long-held, *a priori* ideas about the emission of energy and to adopt a radically new theory on the nature of matter.

The new quantum theory of radiation also explained some long-observed but ill-understood spectral phenomena. Although spectral lines were known to be associated with the presence of particular chemical elements, no one had provided a coherent theory that would explain both the number and location of these lines. As a result, it was left to the brilliant ruminations of the young Danish physicist Niels Bohr (1885-1962) to provide another key to our understanding of the quantum theory of matter. These ideas came from the study of the structure of the atom.

While in Cambridge, Bohr was working towards a coherent theory of the structure of the hydrogen atom. He had accepted the model of the atom revealed by Ernest Rutherford. The classical physics of Scotsman James Clerk Maxwell (1831-79), however, held that the electrons must orbit the nucleus in order to maintain stability. Moreover, Maxwell's physics also predicted that as electrons were accelerating charged particles, they would have to emit energy in the form of electromagnetic waves. Given a configuration analogous to a miniature solar system, the orbiting 'planetary' electrons would be expected to lose their electrical and motional energy through this radiation of electromagnetic waves and, since staying in orbit requires energy, to eventually collapse toward the nucleus, thus putting an end to physical reality. This, of course, does not happen. Yet again the mechanics of Newton failed to account for an irrefutable truth of nature.

In 1913, Bohr proposed that to account for the stability of electrons in orbit around the nucleus of the hydrogen atom, electrons must only occupy *certain* energies, each energy corresponding to a particular orbit. Thus, orbits could be regarded as 'energy levels'. In going from a higher energy level to a lower one, the electron must lose a packet of energy corresponding to the difference in energy between the two orbits. This energy packet is released as a photon of light.

Bohr's brilliant new theory could explain why hydrogen atoms absorb light only at specific wavelengths. The mysterious lines in the solar spectrum that so mesmerised Fraunhofer, Bunsen, Kirchoff and Balmer could now be explained in terms of these new quantum ideas. It was clear that quantum theory was far more effective and far-reaching than any classical model could be, and it was to reveal even stranger truths about the nature of physical reality, as we shall delve into a little later.

Although the Bohr model described the spectral properties of the hydrogen atom with extraordinary precision, it failed to provide a good explanation for the spectral signatures of heavier atoms. This frustrated many physicists, but luckily, a flood of new experimental results on the nature of matter led to a steady refinement of the mechanics of the atom.

In 1902 the German physicist, Philipp Lenard (1862-1947) had reported his experiments on the *photoelectric effect*. Suppose you shine light on a well-polished metal surface. The photons of light carry energy, and are capable of ejecting electrons from the metal surface. These mobile electrons produce a measurable electric current. Lenard's experiments showed that particular metals would eject electrons only if the light used had a high enough frequency. What's more, he demonstrated that if for a given metal the light used was below a certain *threshold frequency*, no electric current could be measured. With some metals, red light, no matter how intense the source, could not induce a measurable current, while blue light did the job splendidly. The early quantum ideas of Planck showed that the energy of a photon is directly related to its frequency. Red light has a lower frequency than blue light, and so will have less inherent energy.

Lenard showed that the production of a measurable photoelectric effect did not depend on the intensity of the light subjected to the metal plate. He also reported that there was no apparent time delay between the arrival of the light at the metal surface and the emission of electrons. These curious results were studied by Einstein, who was forced to conclude that light behaved like miniature bullets, arriving in little packets, or quanta, of energy. This is precisely what one would expect if the light-packets (photons) were particle-like.

In the 1920s the American physicist Arthur Holly Compton (1892-1962) revealed still more startling properties of light. In a famous experiment, he directed high-energy X-rays along a path ending with a metal. He then measured the angles through which the X-rays were scattered, as well as their energies. His results showed that the scattered X-rays had longer wavelengths than the incident rays. This had to mean that they lost energy in the collision, just as if they were particles. In particular, this *Compton effect* suggested that photons have a property called *momentum*. According to classical physics, momentum is simply the product of an object's mass and velocity. In the Compton effect, the X-rays were seen to lose momentum. But contemporary physics insisted that light waves or any other form of electromagnetic radiation had no mass, and therefore, no momentum. Photons, it seems, have properties of mass without possessing the implicit quality of mass. Incidentally, this was also the first clue that led Einstein to postulate that energy and matter are one and the same phenomenon. The particle-like properties of light waves were a contradiction in terms. How could matter be both a wave and a particle at the same time?

Inspired by these experimental findings, the young Prince Louis de Broglie (1892-1987) proposed that since waves have particle-like properties, the converse should also hold. A particle, he asserted, has an associated wavelength, now called its *de Broglie wavelength*. For this bold idea, the young prince was awarded his Ph.D. in 1927, but just by the skin of his teeth! Yet de Broglie's brilliant prediction was actually borne out two years before the publication of his Ph.D., when two scientists, Clinton Davisson and Lester Germer, working at Bell Telephone Laboratories, discovered that electrons produce wave-like diffraction patterns when they collide with metals. In isolation, the photoelectric and Compton effects, as well as the de Broglie hypothesis, were disjointed pieces of a complex jigsaw puzzle. Fortunately, there were enough pieces available to solve the puzzle, but it was to take two great intellects to penetrate this atomic labyrinth. They were Erwin Schrödinger (1887-1961) and Werner Heisenberg (1901-76).

In the late 1920s, Schrödinger and Heisenberg independently studied the new quantum phenomena revealed by Lenard, Compton and de Broglie. They formulated elegant mathematical models to impart both particle and wave-like properties to the electron in the hydrogen atom. The mathematical squiggles of Schrödinger and Heisenberg embody the science of *wave*, or *quantum mechanics*. This new science attested that the behaviour of matter is neither wave-like nor particle-like, but rather both. Where once electrons were conceptualised as tiny impenetrable spheres, quantum theory describes them with a so-called *wave function*. Furthermore, there was an unavoidable uncertainty in particular aspects of atomic behaviour, arising naturally from their quantum models. Let's take a look at just a few.

Uncertainty Rules the World!

Let us perform an experiment in our imagination – a veritable *gedanken* (or thought) experiment of the kind so beloved to Einstein. We set out to measure the exact position of an electron at a definite time as it orbits the nucleus of the atom. Of course, we would have to build suitably refined instruments to try to capture the electron in this particular orbital frame of space and time. We begin to probe the location of the electron. We might perceive it as a piece of ghostly fluff whizzing around the nucleus. As we draw ever nearer to the electron, our capturing device is released and, in the instant of trying to snatch it, the interaction of the instrument with the electron profoundly alters its position. In measuring the position of the electron we unavoidably alter its precise location. As a result, it is not possible to define for certain the location of an electron orbiting the atomic nucleus, but in our attempts to estimate its position, we can only speak in terms of its *probable location*. These ideas were first put forward by Heisenberg and his colleagues and are embodied in the now famous *uncertainty principle*. At its core lies the idea that in our attempts to precisely measure the properties of a particle, like its motional energy or its exact location, we encounter an unavoidable indeterminism, an inherent *graininess* in the mystery of matter. In fact, according to Heisenberg, there is no such thing as an electron that possesses both a precise momentum and a precise position. As Heisenberg said, "We cannot know as a matter of principle, the present in all its details."

According to quantum theory, the concept of a particle is something that is non-localised in space. This is in stark contrast to our common-sense approach, where a particle has both an absolute position and absolute speed. Furthermore, quantum theory tells us that a particle moving in space has a small but finite probability of being located anywhere at any particular time. For example, as we recede from the atomic nucleus, the chances of finding an electron decreases, but never falls to zero. In the hydrogen atom, the probability of finding the electron within 140 picometres (1 picometre=0.000,000,000,001m) of the atom is 90%. So the electron has a 90% chance of being found anywhere within this radial distance from the nucleus. On the other hand, there is a 10% chance that it will be found more distantly than this from the atomic nucleus. As we move farther and farther away from the nucleus, the chance of detecting the electron keeps decreasing – but only reaches zero at infinity. Most physicists would argue that the minuscule probabilities of some of our electrons being located, say, at the distance of the Moon, are of no practical importance. Nonetheless, it is amusing to think that our matter waves permeate the cosmos and that very tiny parts of our being could be everywhere.

Quantum effects get even more bizarre. In particular, the observer plays a fundamental role in deciding the outcome of an event. To gain some insight into the role of observers, consider the simple, imaginary scenario in which a particle can exist in either of two possible locations, A or B. According to our common sense, the particle can be observed at either A or B, but not at both. In this case, quantum mechanics agrees with the Newtonian view. Suppose now we turn our backs for a moment and again ask the question, where is the particle? This time quantum theory asserts that the particle is *both* at A and at B during the time it is unobserved. The unobserved wave functions associated with the particle represent a blending, or *superposition* of both quantum states. It really does exist at two different places at once! But by turning around again, we cause the collapse of the composite wave function, condemning us to observe a definite outcome – the object will be either at location A or at B, but not both. We humans, by virtue of our minds, have the ability to descramble wave functions so that we arrive at a definite answer. The human mind is a kind of wave function descrambling device!

There are still more mind-boggling conclusions that quantum mechanics can draw about the nature of the world. For example, quantum theory asserts that there is no state of rest. To be at rest, a particle's position and velocity (zero in this case) would be known simultaneously. As we have seen, the uncertainty principle predicts that it is not possible to know, simultaneously, a particle's speed and position. An object that we observe as completely motionless is and always will be in motion! It also follows from this argument that an observer cannot know a particle's definite path. We are permitted to have a fuzzy idea of where the particle is heading, but are denied knowledge of its precise path.

Perhaps the most bizarre effect permitted by quantum physics is to allow a particle to exist in a region that is classically forbidden. Suppose we imagine a particle confined to a deep well. Classical physics says that the particle will be retained within the well so long as it doesn't gain enough energy to leave it. But quantum theory predicts that there is a distinct possibility that the particle will be found outside the well even though it may not have enough classical energy to do so! Particles can tunnel their way, quantum mechanically speaking, through an energy barrier. This bizarre phenomenon is called *quantum tunnelling*.

While it is easy to dismiss these 'silly' quantum ideas in our everyday world, we must remember that without these quantum laws, the Sun and the other stars would not shine, the semiconductor and the electron microscope could not have been invented, and lasers would not be around. Quantum mechanics provides the bedrock upon which much of our high-technology is based.

The Dynamic Cosmos

The revolutions in physics heralded by the science of quantum theory allowed astronomers to begin to piece together the nature of the universe. Through spectroscopy we have discovered another great truth of nature: the recession of the galaxies. A good, but not entirely accurate analogy for this recession, can be found by studying a wave phenomenon called the *Doppler effect*. In 1842, the Austrian physicist Christian Johann Doppler (1803-53) suggested that sound waves from a moving object are compressed in the direction of movement and elongated as the object recedes.

Imagine that you are hearing the whistle of an approaching train. As it heads towards you, its pitch increases and as it recedes its pitch decreases. The change in pitch is due simply to the fact that the number of waves striking your eardrums per second changes because of a compression of waves in the direction of motion and an expansion of the same waves as the train moves away from you. Now imagine the same thing for light. Think of a luminous object fleeing away from you at a colossal speed. If it has a spectrum, then it is possible to determine the speed of that object by measuring its Doppler shift, as Doppler himself advocated. Light waves moving towards us will be compressed and will therefore be shifted toward the blue end of the spectrum. If, on the other hand, the object is receding from us then they are elongated, or spread out, causing the light waves to be shifted to the red end of the spectrum. Pitch is to sound as colour is to light.

In 1929, the American astronomers Edwin Hubble (1889-1953) and Milton Humason showed that the distant galaxies have enormous red shifts, consistent with the idea that they are fleeing away from us. But the origin of the red shifts observed for the distant galaxies is not due to the Doppler effect. Instead, it is attributed to the fact that space itself is expanding. The stretching of space causes the rays of light from distant galaxies to cover greater and greater distances. As a result, the light waves become elongated, increasing in overall wavelength. But if everything is fleeing apart then why don't we observe the galaxies breaking up? The answer is due to gravity. The force of gravity glues the 100 billion or so stars of a galaxy together, causing it to move as a unit with the expansion of space.

In the middle of the nineteenth century, Maxwell formulated an elegant and all-encompassing theory of *electromagnetism*. His mathematical work in this field eloquently expressed the findings of the great experimental scientist, Michael Faraday (1791-1867) who had painstakingly worked out the empirical basis of the behaviour of electric and magnetic fields. Maxwell's equations, advanced in 1864, described the intimate relationship between electricity and magnetism and demonstrated that where an electric field exists, a magnetic field is produced at right angles to it, and vice versa.

Such was the power of his work that he predicted the existence of an extensive band of electromagnetic radiation stretching far beyond the infrared region of the spectrum. These waves were purported to have very long wavelengths with low frequencies. Maxwell thus predicted radio waves. But the fruits of Maxwell's seminal work were not to be harvested for some twenty-five years after his electromagnetic equations first appeared. In 1888, the German physicist Heinrich Rudolf Hertz (1857-94) generated an oscillating current from a metal coil, originally devised by Faraday. This device could not only produce, but could also detect radiation with extremely long wavelengths. These were the first radio waves consciously detected and transmitted by a human being. This technology was to blossom, slowly at first, but eventually it would provide another magic eye with which astronomers could study the cosmos.

A young radio engineer, Karl Jansky (1905-50) is largely credited as the first person to use radio waves in applications allied to astronomy. In 1931, Jansky was working with the Bell Telephone Laboratories. His job was to investigate and alleviate the static interference that always accompanied radio-communication. Quite by accident, he noticed that when he pointed a radio receiver at the sky, he always detected a hissing noise. At first, he attributed these radio sources to some unknown terrestrial source, but after two years of meticulous investigation, he was certain that they must come from the Milky Way. What's more, the radio noise was greatest in the direction of the galactic centre in the constellation of Sagittarius. These serendipitous events marked the beginnings of the new science of radio astronomy.

Since radio waves are just another form of electromagnetic radiation they can be collected and brought to a focus, much like ordinary visible light, so that a radio-image of the object can be constructed. Compared with optical telescopes, radio telescopes need to be much larger in order to focus the longer wavelength radio waves. While the largest optical telescopes on Earth are of the order of 10 metres in diameter, the largest in-ground radio telescope in the world, located at Arecibo, Puerto Rico, has a diameter of 305 metres! Today, our radio telescopes are peering deep inside stars, galaxies and, closer to home, hot radio objects such as the infernal surface of Venus and Jupiter's deep interior.

Similar use has been made of infrared (IR) light. IR radiation conveys a lot of information on the temperature of an object, as well as a surprising amount of detail about the chemical properties of matter. IR radiation excites molecules and causes them to bend, stretch or oscillate. These movements betray themselves to us through the production of characteristic IR spectra. By studying the peaks and troughs of these spectra, astrophysicists can identify many large and small molecules even in the near-perfect vacuum of interstellar space, including a whole host of carbon-based substances – molecules that are identical to those found in living things on Earth.

The opportunities created by our species in opening up the windows of the electromagnetic spectrum are incalculable. We now have instruments that can detect radiation from almost all accessible parts of this spectrum. With the aid of optical, infrared, microwave, radio, ultraviolet, X-ray and gamma-ray astronomy, the human mind has penetrated deep into the heart of nature. And with these new eyes we are now prepared to venture back through space and time to the epoch of the cosmic dawn. Let us regress into the provinces of our imagination, where we will begin to retrace some of the events in the history of the beautiful cosmos from which we spring.

Creation *Ex Nihilo*?

In the beginning there was nothing. This seems hard to grapple with. How can something come from nothing? One way round this conundrum is to revise what we really mean by nothing. Let's look at the concept of zero for a moment. Imagine that you are fortunate enough to have £100,000 in the bank. You are substantially in the black! Now suppose that you owe someone £200,000 pounds and you pay your debts. Your exasperated and bewildered bank manager would not allow the cheque to be paid as your account would then be £100,000 overdrawn. Which would you prefer: zero pounds in the bank or £100,000 in the red? While your creditor might not be delighted, you and your bank manager would feel happier. I rest my case! Clearly in terms of numbers, zero lies precisely halfway between minus infinity and plus infinity.

Most people have a hard time visualising nothing. The nearest I can get to imagining it is to picture a vacuum. Yet quantum theory sees the vacuum as something else entirely. According to particle physicists, a vacuum is a flat field, where particles and their corresponding anti-particles flutter in and out of existence. Physicists refer to these as *virtual pairs*. On the vast majority of occasions, the virtual particle and its corresponding anti-particle collide and annihilate each other causing no net perturbation of the vacuum. But if this vacuum is disturbed, spectacular events can occur.

Our universe, we are told, came into being when virtual pairs of particles inflated the vacuum field, draining it of energy and creating matter in the process. The vast and ancient universe that we live in may well have been the result of an extraordinary runaway inflationary event. Scientists refer to this bizarre and poorly understood phenomenon as a *quantum fluctuation*. If this is in some sense what happened (and there are some that have expressed their doubts), we might well expect other universes to be popping in and out of existence as you sit and read this book! Most may have fleeting existences, coming into being for only the tiniest fraction of a second, while others may be older and far more exotic than the universe we are condemned to experience.

The idea that all matter and energy was created in the great primordial fireball of the Big Bang is hard to grapple with. But as Einstein taught us, energy and matter are the same. Moreover, energy can exist in two different forms – positive and negative. The positive energy is locked up in the various forms of electromagnetic radiation and kinetic (motional) energy of the cosmos. Most of the negative energy is bound up by gravitation. When we add up the total amounts of negative and positive energy in the universe, a strong case can be made for asserting that the universe has zero energy! Negative and positive energies cancel each other out. What's more, there is exactly the same number of positively charged particles in the universe as there are negatively charged particles.

Particles also have an inherent spin, which can be either clockwise or anti-clockwise. When the sums are done on the net spin of all the matter in the universe, it too comes up with zero. Everything that exists, every person, flower, planet, star and galaxy, came from the quantum vacuum, something very close to, but not quite, nothing. Yet to my mind, the vacuum field must have had the right properties to allow a four-dimensional space-time to become manifest. Where did that come from? Alas, it is here that we must bow to mystery. Nobody knows.

Boom!

Before the Big Bang, *nothing* existed. No matter, light, space, nor even time. A quantum fluctuation caused a point of energy to come into exist-ence and from it sprang forth everything that was, is, or ever shall be. From this entity, smaller than a subatomic particle, all the matter in the universe was derived. There was a great explosion, leading to the expan-sion of the fabric of space, the conversion of energy into matter, the synthesis of atoms, and eventually stars, planets and presiding over it all, contemplating and curious, life. The rapid, inexorable expansion contin-ued, and with it a great cooling occurred. Within one hundred billionth of a second subatomic particles, known as *quarks*, came into existence. Within one ten billionth of a second after the Big Bang, quarks embraced each other to form the familiar subatomic particles, the protons, elec-trons and neutrons. Between one and five minutes after the birth of the cosmos, these particles interacted to form the simplest atoms, hydrogen and helium and a little bit of lithium thrown in for good measure. From these simple chemical elements, all others can be formed.

How do we know that the universe began as a Big Bang? We have touched on some of the evidence. The observed redshift of the distant galaxies tells us that they are fleeing away from us in all directions. And because the redshift increases as the distance between the galaxies increases, so too must the speed of recession increase commensurably. This idea, as we have seen, was first described by Edwin Hubble and Milton Humason in 1929. Indeed they showed that there was a simple law relat-ing the speed of recession of the galaxies to their distances from the Earth. Today, we call it, appropriately enough, *Hubble's law*. Here it is:

$$V = H_0 \times R$$

Where V = velocity of recession (in kilometres per second)

R = Distance in Megaparsecs (Mega represents 1 million. One parsec is a little over three light years)

H_0 = the Hubble constant (in kilometres per second per Megaparsec).

If we were to plot this relationship as a graph, we would get a straight line with a gradient equal to the Hubble constant. With a reliable value of H_0 and knowledge of the amount of matter in the universe, we can answer two profound cosmological questions. What is the age of our universe, and will it expand for ever?

Although we are still unsure about the answers to these difficult questions, there are reasons to think that we are very near to resolving them. Thanks to science, we know that the single factor that will dictate the future behaviour of our universe is *mass density* – a measure of how tightly matter is distributed in space. Matter, through gravitation, attracts other matter. This gravitational tug, experienced everywhere in the universe, provides a collective force that opposes the expansion. Is there enough matter in the universe to grind the expansion to a halt, perhaps triggering a contraction in the course of events?

There is obviously an abundance of matter in the cosmos. There is the kind we can look at – the shining gaseous nebulosities, the ghostly wisps of light we see as the galaxies. But a careful census of galactic mass-distribution shows that there is not nearly enough matter to account for the motional properties of the galaxies. Maybe there is more matter in the universe than meets the eye?

Illuminating the Invisible

In recent years, astronomers have gathered increasingly convincing evidence that the vast majority (>90%) of the matter in the cosmos is invisible. Our universe, it is believed, is made mostly of a substance or substances we know virtually nothing about. We call it *dark matter*. All we know about this elusive matter is its gravitation.

When astronomers first analysed the dynamics of our Galaxy's motion, they were astonished to discover that a lot of this missing matter seems to be located on its outskirts. It seems that this dark matter is physically separated from the matter that makes up planets, stars and all life forms. Some astronomers think that it could make up trillions of Jupiter-like worlds, or small stars denied the ability to shine in visible light. But the lack of detection of substantial numbers of these stars in the Galaxy most likely precludes their abundant presence in the cosmos.

Most astronomers now believe that the leading contenders for the missing mass are electrically neutral entities known as *neutrinos*. These strangely elusive bodies were, up to recently, widely held to be massless, travelling at the velocity of light. But the results of years of observations by a team of Japanese scientists using a fluid-filled tank called the Super Kamiokande detector, are beginning to throw light on the dark matter problem. Located 1,000 metres below the ground in an old zinc mine, the detector consists of a 12.5 million gallon tank filled with ultrapure water. On the walls of the gigantic detector lie over 13,000 light-detecting devices (photomultiplier tubes). The idea is that a neutrino passing through the Earth will collide with a water molecule and in so doing, emits an extremely brief flash of light. By detecting these light pulses, the scientists have been able to deduce some of the characteristics of neutrinos. What's more, their results strongly suggest that these elusive particles have mass. Indeed, the team has inferred that the prodigious numbers of neutrinos may account for a considerable part of the missing mass of the universe.

Although neutrinos may have little mass, their numbers are truly phenomenal. Every second, trillions upon trillions trace their way from the Sun towards the Earth. But even when the total mass of the neutrinos is taken into account, many astronomers still believe that there is not nearly enough matter to halt the expansion of the universe. We may well be poised forever on a wave of expanding space-time.

Cosmic Yardsticks

Although physicists are confident that the recession of the galaxies, as measured by Doppler shifts, is based on sound physical principles, it is the correct measurement of the tremendous distances to the galaxies that is the subject of heated debate among cosmologists. Astronomers have devised many ingenious methods to measure these colossal distances. For example, a number of stars in the sky undergo periodic variations in luminosity. The story of how these variable stars came to be used for measurement began with the promising young English astronomer, John Goodricke (1765-86). In 1784, just two years before his untimely death at the tender age of 21, he discovered that Delta Cephei, a fairly bright star in the constellation of Cepheus, displayed a regular pattern of brightening and dimming. Astronomers have since found a large number of other stars that vary in the same regular way, and have accordingly classified them as *Cepheid variables*. Cepheids can have periods ranging from a few days to nearly two months.

In 1912, the Harvard astronomer Henrietta Leavitt (1868-1961) was studying two dwarf satellite galaxies of our own, the Large and Small Magellanic Clouds, looking for Cepheids. She found 25 in all and recorded the period of variation of each. To her astonishment, Leavitt noted that the longer the period, the brighter the star. This immediately suggested to her that in principle, Cepheid variables could be used to measure enormous intergalactic distances. By measuring the periods of the nearby Cepheids with high precision, she established a period-luminosity curve, showing the relationship between the period of the Cepheid and the star's intrinsic, or *absolute* luminosity. Her initial discoveries were later refined by Harlow Shapley (1885-1972) and Ejnar Hertzsprung (1873-1967). Using this Cepheid yardstick, Shapley set out to measure the distance to the centre of the Milky Way. He found that the star clouds in the galactic centre were some 50,000 light years distant. This figure, although about 50% larger than modern estimates, nonetheless demonstrated the immense distances between our location and the centre of the Milky Way.

Since the introduction of Cepheids into galactic astronomy, astronomers have expanded their portfolio of *standard candles,* or objects of known luminosity embedded within objects whose luminosity one wants to measure. Some are based on the properties of stars and their ghostly remains, while others are based on the properties of whole galaxies. Refining distance estimates for distant galaxies is crucially important if we are to understand how our universe came into being and what its fate may be.

Giant Catherine Wheels in Space!

Many people are aware that we live within the confines of a spiral galaxy. But how can we deduce this if the Sun and its family of worlds are embedded deep within it? The first attempt to elucidate our location within the Milky Way was made by the celebrated astronomer of antiquity, William Herschel, who in the 1780s counted the number of stars in each of 683 areas of the sky. Herschel found approximately the same numbers of stars all over the Milky Way, and concluded that our position must be near its centre. Like a giant amoeba with a nucleus at its centre, Herschel once again gave our Sun a special position at the nerve-centre of the known universe. The Herschelian view prevailed right up to the early 1900s.

Yet Herschel never knew that dust obscures starlight from very distant objects. The amount of light dimming depends on the concentration of interstellar dust, as well as the wavelength of the light penetrating it. Electromagnetic radiation with relatively short wavelengths, such as visible and ultraviolet light, interacts much more strongly with interstellar dust than does longer-wavelength radiation such as infrared and radio waves. For this reason, scientists using radio and infrared telescopes can penetrate much more deeply into the elusive dust lanes of the Milky Way. Astounding advances in our knowledge of the Sun's location within the Galaxy were achieved through the work of the Infrared Astronomical Satellite (IRAS) in the 1970s, and more recently by the Cosmic Background Explorer (COBE). These spacecraft have probed the distribution of interstellar dust at long and short infrared wavelengths and have revealed the long, slender structure of the Milky Way, together with a large, bright central bulge, suggestive of a spiral arrangement viewed edge-on.

In the 1920s, Harlow Shapley mapped the three-dimensional positions of a large number of globular star clusters – spherical collections of up to a million stars tightly bound-up by gravity – located above the mid-plane of the Milky Way. He found to his surprise that they were not centred around the vicinity of the Sun, but seemed to trace out a huge, spherical shell centred on the star clouds in the constellation of Sagittarius. Shapley boldly conjectured that the Sun was located far out from the core of the Milky Way Galaxy and that the star clouds in Sagittarius marked the direction to its centre.

Perhaps the most elegant work demonstrating that our solar system orbits the galactic centre comes from the study of Doppler shifts of stars located at increasing distances from the Earth. Because radio waves can easily penetrate the obscuring dust in interstellar space, they can be used to deduce the relative velocities of stellar groups located at these various locations within the Galaxy. Picture our Galaxy with a number of arms spiralling out from its central bulge. Stars located on spiral arms between us and the galactic centre should move at higher orbital speeds than those in a spiral arm farther out from the Galaxy's centre than the Sun. In similar fashion, groups of stars located at or about the same distance from the galactic centre as our Sun should have a very similar orbital speed to the Sun.

We have already seen how light from a moving object is Doppler shifted, with faster objects exhibiting greater shifts. Radio sources, such as atomic hydrogen, also display Doppler shifts. Using radio astronomy, scientists have confirmed that stars located farther out than the Sun have small Doppler shifts, while those located nearer the galactic centre than the Sun exhibit larger Doppler shifts. Moreover, radio astronomy mapping has revealed that stellar groups are not strewn evenly across the Galaxy, but seem to be condensed in discrete segments. These observations are all consistent with the idea of a spiral-structured stellar metropolis. In fact, we now know that the Sun is located about 25,000 light years from the centre of the Milky Way Galaxy and takes about 200 million years to orbit its nucleus.

We have seen that the galaxies appear to be fleeing away from us and that their recessional speeds are proportional to their distances from our Galaxy. This seems to indicate that our universe is expanding. If we were to run the cosmic movie further and further back in time we would observe a smaller and smaller universe. Eventually, at some moment about 12 billion years ago, the universe would be seen to occupy no space at all. It would disappear into an infinitely small space, or as astronomers call it, a *singularity*. We would have travelled back to the epoch of the Big Bang. In an expanding universe the density of matter decreases, but if we were to run the clock of time backwards, its density would have to increase. At the instant of the Big Bang, the matter in the primordial cosmos must have been very hot and dense. The super-intense radiation from an event of this magnitude would send reverberations down through the aeons, preserved even in our epoch.

Echoes of Creation

As the universe expanded, the matter therein cooled. And just as an experienced steel worker can estimate the temperature of a furnace from the colour of light it emits, astronomers have picked up the cooled radiation from cosmic embers 12 billion years old. In 1949, the physicist George Gamow (1904-68) demonstrated, theoretically at least, that the radiation accompanying the advent of the Big Bang should have lost energy as the universe expanded, and would now exist in the form of radio-length waves coming from all parts of the sky. He calculated that the temperature of this *background radiation* should have cooled to a temperature just five degrees above absolute zero. In May 1964, two astronomers, Arno Allan Penzias (b. 1938) and Robert Wilson (b. 1936), detected a radio wave background with characteristics strikingly similar to those predicted by Gamow. The COBE satellite has confirmed the general uniformity of this background radiation in an impressive, all-sky survey completed in the early 1990s. Our species has unearthed the fossilised outcries of a great and ancient cosmic creation. Through the discovery of the recession of the galaxies and the cosmic background radiation, we have strong reasons for believing that our universe originated in a tumultuous explosion.

Although the COBE survey demonstrated the general uniformity of the cosmic background radiation, it also showed that on a very fine scale the radiation was 'lumpy'. This came as a relief to the astronomical community because it showed that matter has an innate tendency to aggregate, thus providing scientists with an explanation of how the galaxies could start to form. The first steps in galaxy formation could not have been taken if the expansion had produced a completely uniform distribution of matter and energy.

There are even more compelling reasons for believing in a Big Bang of some sort. Physicists have constructed computer models of the early universe on the assumption that it began as an infinitesimally small point and inflated outwards at a prodigious rate. These models predict that the chemical composition of the cosmos should be about 75% hydrogen, with the vast majority of the remainder appearing as helium. These are exactly the proportions recorded by astronomers. For now at least, the Big Bang theory of the origin of the universe remains by far the best explanation for the events that led to the grand and imposing universe we inhabit today.

Whence the Shining Stars?

The most recognisable entities in a galaxy are stars. How do they form? Astronomers have amounted a formidable edifice of knowledge showing that stars are born when cool, giant clouds of hydrogen and helium, peppered with dust grains and simple molecules, collapse under their own weight. The impetus for this contraction is provided by gravity. Eventually, a giant molecular cloud fragments into a series of dense clumps. The matter at the centre of each clump gets compressed, with the result that the core becomes super-hot – so hot that thermonuclear reactions are 'kick-started'. Gradually, the clump flattens into a thin, spinning disc of matter, which eventually forms a family of planets circling a central star. The cosmos proclaims the nativity of a solar system. This is more or less how our star, the Sun and her attending worlds were spawned from a nebulous haze of gas and dust about 4.5 billion years ago. But the refulgent stars do magical things. Inside stars, the temperatures are high enough to promote the transmutation of one element into another.

The Big Bang gave rise only to the two simplest elements, hydrogen and helium, with perhaps some lithium thrown in for good measure. Yet we now know of the existence of more than 90 naturally occurring elements. Where did iron and lead, silicon and carbon, gold and uranium come from? Might they have somehow 'evolved' from hydrogen and helium? All atoms have a dense nucleus, made up of positively charged particles called protons and neutral particles named neutrons. The nucleus contains the vast majority of the mass of the atom. But circling the nucleus are the tiny, negatively charged electrons, possessing only about one two-thousandth the mass of the proton. At very high temperatures (about 15 million degrees), such as those attained deep inside stars like the Sun, the electrons are excited so much that they break free from the electrical embrace of the nucleus forming *ionized* atoms or *plasma*. Because heat imparts kinetic energy to particles, the resulting isolated nuclei are free to smash into one another. Occasionally, the collisions are suitably energetic to fuse two smaller nuclei into one heavier nucleus. This process is called *nuclear fusion*, and it liberates enormous amounts of energy.

Four hydrogen nuclei can be transformed into a single helium nucleus by nuclear fusion. It is this 'hydrogen burning' that powers over 90% of the stars in the Galaxy. This stable phase in the life of a star is called its *main sequence*. In more highly evolved stars, other reactions are possible. Three helium nuclei can combine to form a single heavier nucleus, carbon. By colliding in different ways in the searing interiors of stars, hydrogen and helium nuclei can spawn heavier elements like oxygen, beryllium, nitrogen and iron atoms. It is from these atoms that life on Earth took its first step.

The Extraordinary Lives of Stars

Like human beings, stars have their own lives. They are born, they live and then they die. But unlike human lives, the legacy of stellar life is measured not in tens of years, but in millions or billions of years. And as the chronicle of a human life grows and blossoms like a flower, so too it is for the stars. And the comparisons don't end here. Some massive, brilliant-white stars are so precocious that they burn their entire reserves of hydrogen in only a few million years, while many others shine for hundreds of billions of years. Moreover, just as in a human city, stellar cities contain a mosaic of stars of different ages. For example, many people in the northern hemisphere may have spotted a cluster of stars in the constellation of Taurus, popularly known as the seven sisters, or as astronomers call it, the Pleiades cluster. Located just 400 light years away, the beautiful light of these brilliant blue-white stars is veiled in a ghostly wisp of bluish nebulosity, the remainders of the gas and dust that formed the stars in the cluster long ago. The stars making up this so-called open cluster are only about 50 million years old and are just beginning to venture out of their stellar nursery.

Our Sun, in comparison, is a middle-aged star, well past the uncertain time of its youth. It has already gone through 4.5 billion years of life and will continue to shine salubriously for at least a few billion more. Meanwhile, in Hercules, a summer constellation in the northern hemisphere, a pair of binoculars will show you a small, spherical collection of stars bound together by gravity: these are called the globular clusters. Although some are embedded in the disc of the Galaxy, many globular clusters surround the Milky Way, like a swarm of bees around a hive. The stars making up these beautiful clusters of more than 100,000 suns represent some of the oldest stars in the universe. In fact, astronomers believe them to be approximately ten billion years old.

The stars are like us. Some live fast, furious lives, promptly destroying themselves in only a few million years, while others live life at a slow, relaxed pace, ending peacefully in tens or hundreds of billions of years at a grand old age. A star can end its life in a variety of ways. The critical factor dictating the manner in which a star ends is its mass at birth. After the Sun, a medium-sized star, has converted all its hydrogen into helium, it will use the latter as a fuel to prolong its life span. Helium nuclei will fuse to form heavier atoms, mostly carbon and oxygen. As it burns helium in its core, it will expand and brighten, becoming a red giant hundreds of times more luminous than it is now. Its inexorable expansion will be so awesome that it will engulf the inner planets of the solar system. The rocky worlds of Mercury and Venus, and possibly the Earth too, will meet their demise in this way.

The core of the Sun will fill up with oxygen and carbon nuclei and it will shed a vast expanse of matter from its outer layers into interstellar space. It will end its life as a small, hot, blue core, hardly the size of the Earth, surrounded by luminous stellar ejecta: a so-called *planetary nebula*. After shedding its outer layers into interstellar space, the Sun will slowly dim and fade over a period of several hundred million years. What will remain will be a burnt-out stellar corpse of very high density – a *white dwarf*.

In contrast, stars born with greater than about 8 solar masses end their lives in a final cataclysmic fireworks display. High-mass stars tend to be hot and energetic, squandering their reserves of hydrogen and converting it into helium. They become *red supergiants*. You can see several bright red supergiant stars in the night sky. Betelgeuse, in the constellation of Orion, and Antares, in Scorpio, are good examples. Red supergiants burn helium into carbon and oxygen as lower-mass stars do, but the former then undergo further rounds of nuclear fusion, creating still heavier atomic nuclei such as magnesium, sulphur and silicon.

But something strange happens after the star synthesises silicon – it sounds the death knell for these beautiful giant stars. Silicon gets converted into iron, and iron does not burn (in a nuclear sense) at all. Because the synthesis of atoms with masses greater than iron requires more energy than they release by undergoing fusion, a crisis ensues; the star collapses under gravity and implodes, becoming a *supernova*, shining briefly with the luminosity of a billion stars. Only the core of the star remains after its violent death. The electrons and protons making up the matter in the supernova remnant combine to form neutrons, releasing a torrent of neutrinos in the process. The remnant is like a giant atomic nucleus – a *neutron star*. And just as a pirouetting skater spins more rapidly as she draws her arms inward, the tiny stellar core – only a few tens of kilometres across – speeds up its rate of rotation. While stars like the Sun spin on their axis in times measured in days, neutron stars may complete their axial rotation in fractions of a second. These bizarre stellar corpses are what astronomers call *pulsars*. The fastest pulsars (so-called millisecond pulsars) revolve at speeds of about 10% the speed of light. Pulsars give off extraordinarily regular pulses of radio waves and other forms of electromagnetic radiation – flashing on and off like a lighthouse many times every second.

But neutron stars can't be any old size. There is a limit to the amount of matter a neutron star can possess before gravity gets the upper hand. If the burnt-out core from a supernova exceeds about 2.0 solar masses then nothing can stop gravity crushing it out of existence. The star is doomed to end its life as a black hole – a place where a star once was.

The Mysterious Black Hole

Suppose we consider a massive object like the Earth. As we all know an object thrown upwards will always fall down. This is because the Earth exerts a gravitational force centred at its core, which pulls the object back to Earth. But if we give a rocket or a missile enough kinetic energy, it can break free from the gravitational influence of the Earth and escape into space. The speed required for an object to escape the Earth's gravity is known as its *escape speed* – about 11.2 kilometres per second.

Suppose now we steadily crushed the Earth down from its present radius of 6,378 kilometres to smaller and smaller radii. As we continue to crush the Earth, its gravity gets more concentrated, increasing the escape velocity to higher and higher values. But as we have seen, the speed of light is finite. How much would we need to crush the Earth before light could not escape? The answer – to a little under 9 millimetres. To make a black hole out of the Earth we would need to crush it down from its present radius of 6,387 kilometres to just 9 millimetres!

Contrary to popular belief, the idea of a black hole is not a product of twentieth-century physics but rather of the eighteenth century. In the fateful year of 1783, the English clergyman John Michell (1724-93) examined the consequence of gravity on the escape of light from an object like the Sun. Using Newton's theory, he calculated thus: "If there really should exist in nature any bodies, whose density is not less than that of the Sun, and whose diameter is more than 500 times the diameter of the Sun ... their light could not arrive at us." This startling leap of imagination went largely unnoticed until the great French mathematician and astronomer, Pierre Simon Laplace (1749-1827) rediscovered the idea in the late 1790s. Nonetheless, the idea that an object might exist that was sufficiently massive or dense as to prevent light escaping from it was to remain in the realm of conjecture for over a century, until the advent of Einstein's epochal theories of relativity in 1915. In the 1970s, Stephen Hawking (b. 1942) at Cambridge University, and Roger Penrose (b. 1931) at Oxford University joined forces to show that, according to the general theory of relativity, a black hole must end with an infinitely small and dense region of space-time – a place where the laws of physics as we know them break down.

As well as the relatively small black holes resulting from the collapse of a single, high-mass star, we also know that gigantic black holes may lurk at the centres of spiral galaxies. Astronomers have unearthed strong evidence that a black hole with an estimated mass of over 3 million times that of the Sun is purring away at the centre of the Milky Way. Indeed, astronomers studying the dim and distant galaxies have revealed a host of *active galaxies*, which have as their common denominator a super-massive black hole at their centres. To get an idea of how concentrated matter is in the vicinity of these enormous black holes, imagine stuffing over a billion suns into a space less than the size of the solar system. The matter falling into these enormous beasts of nature releases incredible amounts of energy. Moreover, these super-massive black holes may provide the energy to power the ancient and distant quasars, X-ray galaxies and radio galaxies catalogued by astronomers over the last few decades.

The Cosmogony of Worlds

Fortunately for us, not all the matter in the universe condensed to form the stars and black holes. In addition, myriad worlds were spawned. As we have seen, stars are formed from the top down, collapsing from a large molecular cloud of gas and dust. Planets, on the other hand, are formed from the bottom up, by sweeping up gas and dust from the disc of matter encircling the star. Over a few million years, planets grew from kernels of matter.

In our solar system, the terrestrial planets – Mercury, Venus, Earth and Mars – are the four innermost worlds and are composed mostly of rock and iron. They are dense planets that have solid surfaces, volcanic activity, and in one place at least, life. Farther out in the cold nether world of the outer solar system, we come to the giant gas planets – Jupiter, Saturn,

Uranus and Neptune. Unlike the terrestrial planets, these worlds are characterised by their greater size and gassy composition. They are generally thought to be relatively pristine, miniature replicas of the Sun, composed of mostly hydrogen and helium with smaller amounts of methane and ammonia, as well as a rich inventory of more complex, carbon-based, organic constituents. Unlike the Sun however, the cores of the gas giants may hold rock and ice totalling several times the mass of the Earth.

Jupiter is by far the most massive planet in the solar system. 1,300 Earths would fit into its vast globe. Early radio studies of Jupiter revealed that it was radiating about twice as much heat back into space as it was receiving from the Sun. This had to mean that its interior was very hot. In fact, had Jupiter accreted only about 20 times more mass than it actually did, the internal pressures would be high enough to kick-start primitive thermonuclear reactions in its core, allowing it to shine feebly by its own light. Jupiter is indeed a star that failed. Imagine the consequences for us had Jupiter become a small companion star to the Sun. Our neck of the cosmic woods would have become a *binary star system.*

Both stars in a binary system orbit around a common centre of gravity. If both stars have equal mass, the centre of gravity will be equidistant between them. If, however, one star is more massive than its companion, then the centre of gravity will lie closer to the bigger star. Binaries and multiple star systems are exceedingly common in the universe. Astronomers estimate that as many as 60% of stars in the Milky Way Galaxy are born as part of binary or multiple star systems. But binary systems spell out bad news for planets. This is because it is far more difficult for planets formed in binary systems to attain stable orbits. However, recent computer programs designed by Robert Harrington at the US Naval Observatory have revealed that, hypothetically at least, planets inhabiting multiple star systems may have surprisingly stable orbits.

Using our solar system as a model, Harrington replaced the Sun with two stars, each only half the mass of the Sun and separated by about 65 million kilometres. Of all the planets, Mercury was the only one that seemed to be much affected by this change. The orbit of Venus suffered only a minor change, while the Earth and the outer worlds of our solar system seemed to hurtle on regardless. Harrington also tried replacing Jupiter with a substitute star of similar mass to the Sun. These simulations also suggested that such a change would not affect the Earth's orbit significantly. How I wonder what it would be like to live on a world with two or more beautiful suns looming in exotic, alien skies. There must be many million of such worlds in the Milky Way Galaxy alone, patiently awaiting an onlooker. Which intelligent species of the Galaxy will set eyes on them first?

A Vast Menagerie of Planets

For many centuries astronomers have speculated on the possibility that other stars had their own family of planets. If the Sun, an ordinary star, possesses a family of worlds, why should it be any different for the distant stars? Perhaps there are a billion Earths in the Milky Way Galaxy alone. As intuitive as our thinking has been, we did not have formal confirmation of the existence of extra-solar planets – worlds orbiting other stars – until comparatively recently. Indeed, it was not until the mid-1990s that astronomers first obtained compelling evidence for the plurality of worlds. Thanks to enormous advances in optical and radio engineering technology, astronomers now know that there exists a great number of extra-solar planets than planets in our own solar system.

How can we possibly detect a planet orbiting a distant star? To gain an appreciation for the magnitude of the problem, let's take the time to assess the brightness of the stars. The great third-century Greek astronomer Aristarchus of Samos first catalogued the positions and brightness of over 800 stars in the heavens. The brightest stars were assigned magnitude 1, while the dimmest stars visible to the naked eye were given a magnitude of 6. A first magnitude star is about 2.5 times brighter than a second magnitude star, and a second magnitude star is about 2.5 times brighter than a third magnitude star. It can easily be shown that there is a factor of 100 separating the brightness of a first-magnitude star and a sixth-magnitude star.

At the distance of Alpha Centauri, located about 4.3 light years distant, the Sun would have an apparent magnitude of 1 and Jupiter would be magnitude 23. In visible light, the Sun shines about a billion times more brightly than Jupiter. But if we shift our vision to the infrared region of the EM spectrum, we can close this enormous brightness gap between Jupiter and the Sun. The Sun shines about 100,000 times less brightly in the infrared compared with Jupiter. This is because the heat generated in the interior of Jupiter causes it to emit about twice as much energy as it gets from the Sun.

Modern large-aperture telescopes, either ground-based or in Earth orbit, are capable of picking up Jupiter at this distance – but it's not as simple as that. Just as the bright sunshine on a summer's day prevents you from seeing the details of your surroundings, so too does the glare from the central star blot out the comparatively feeble light from the planet. To remove the problem of glare, it is necessary to construct telescopes that are capable of blotting out the light from the central star, rendering its attending planets more visible. This technique has been used in the study of the Sun's outer atmosphere, the solar *corona*, for many years and is appropriately known as *coronography*. Such a facility has been equipped on the newly-refurbished Hubble Space Telescope (HST) and will allow astronomers to detect very faint dwarf stars and even large extra-solar planets.

But perhaps the easiest way to infer the presence of planets orbiting nearby stars is to use the tried and trusted technique of *astrometry* – a system of methods which measures tiny changes in the position of an astronomical object over a period of time. This method was first used by the German accountant-turned astronomer, Friedrich Wilhelm Bessel (1784-1846), who achieved the monumental goal of measuring the *proper motion* of the nearby star 61 Cygni, enabling him to estimate its distance. This technique can be applied to the identification of planets because if a star has no attendant planets then it should move in a more or less straight line, while the orbital motion of one or more planets around a star will cause it to wobble as it moves across the sky. These 'wobbles' can be picked up and the masses of the unseen planets inferred.

Alternatively, one may study the orbital motion of the parent stars themselves. It is not strictly true that a planet orbits about a stationary star. More precisely, both the star and the planet orbit about their common centre of mass. This is called the *barycentre*. Astronomers can carefully watch nearby stars and catalogue their position with exquisite precision over a period of time. From the observed motion and period of movement of the star, it is possible to deduce the mass and distance of the planet from the star.

Another adaptation of this approach involves the search for changes in the radial velocity of the star as it moves around its orbit. For example, the Sun changes its velocity by about 13 metres per second over a period of twelve years, owing to the gravitational force of Jupiter. Had Jupiter been located nearer the Sun, the velocity differences would be larger. Velocity differences can be worked out to high precision by working out the tiny Doppler shifts of the star's light as it orbits its centre of mass.

Other methods have been employed to detect *planetary transits* – at the point where a world in orbit crosses the disc of a star, a small dip in the star's luminosity can be registered. By studying these changing patterns, astronomers can glean important, albeit basic, information about the mass and orbital characteristics of these alien worlds.

Most of the worlds so far discovered have masses about the same as that of Jupiter or greater. This is hardly surprising since these planets exert large gravitational effects on their suns, in comparison with the feeble gravitational embrace of small terrestrial planets. But astronomers are making rapid progress in remedying this annoying limitation. In particular, through an ingenious and relatively recent technological advance known as *optical interferometry*, we may soon have the spectacular ability to image details on these far-away worlds, and not just the cloud tops of giant, lifeless Jovian planets, but worlds like Mars and Venus and even other Earths.

Interferometry is based on the addition and subtraction of waves of light. Light waves that are in step (or in phase) with each other can add up. This is called *constructive interference*. On the other hand, light waves that are completely out of step tend to cancel each other out, causing *destructive interference*. In principle, interferometry can be applied to any form of electromagnetic radiation. And although the technique was first pioneered in the 1970s for radio astronomy, there are now welcome signs that interferometry can be modified for use in ordinary, optical telescopes.

Interferometry allows an observer to join several smaller telescopes together in such a way that they behave like a single telescope with an aperture equivalent to the distance between the most distant points in the array. Furthermore, by blotting out the glare of the central star, interferometry may also provide a means of detecting the feeble light of utterly alien worlds – worlds that for now exist only in our imagination.

Interferometry offers unprecedented resolution over older techniques and is destined to revolutionise our view of the universe. Already, plans are underway to track down other Earths. The astronomers of the early twenty-first century will have characterised planetary systems detected around thousands of nearby stars. Moreover, we may soon have the ability to detect the tell-tale stirrings of life on some of these distant planets.

How might we find such life? Many exobiologists believe a good candidate molecule is oxygen. On Earth, the only process that can continually produce oxygen over aeons of time is that pioneered by plants, in an extraordinarily complex mechanism called *photosynthesis*. As we shall see, oxygen is quite a reactive substance. It readily combines with minerals in the Earth's crust causing their oxidation. Indeed, if all of life on Earth were to be callously snuffed out this very minute, the oxygen in the atmosphere would be rapidly depleted in just a few thousand years. So we would expect oxygen to be depleted rapidly on a world that does not continually produce it.

The gas ozone has also been suggested as an indicator of biological activity. However, it is a bit more difficult to substantiate because, while it is of pivotal importance to life on Earth due to its radiation-shielding properties, it is only a very minor constituent of the Earth's atmosphere. Indeed a back-of-a-stamp calculation shows that if all the available ozone in the Earth's atmosphere were compressed down to the surface, it would there form a layer just 3mm thick! In comparison, the major atmospheric gases, nitrogen and oxygen, would form an 'ocean' several metres thick. Thus we do not expect to find ozone easily in the spectrum of light from even an Earth-like extra-solar planet. In detecting ozone on a far-away world orbiting another star, astronomers would be seeing a gas that's barely there at all!

Recent indications are that ozone is present as a trace constituent of the evaporated ices from the surface of Jupiter's moons – Europa and Ganymede. Nevertheless, ozone may turn out to be a red herring in our search for life around other stars in the Milky Way Galaxy. This is because other shielding mechanisms can do the same job, as we will see when we shortly discuss the early evolution of the Earth's atmosphere.

Since terrestrial plants and microbes are responsible for the evolution of oxygen in our atmosphere, it is possible that the presence of this gas would act as a signature of Earth-like biology. Imagine our scientists, meticulously sifting through millions of extra-solar planetary spectra, looking for tell-tale signs of other Earths laden with oxygen. We would have the means to seek out life based on our own chemistry. Similarly, across the great dark of interstellar space, there may be beings who think and wonder as we do, preparing a catalogue of worlds with abundant oxygen atmospheres in anticipation of their own eventual exploration. Maybe this is a way for intelligent life forms to seek out beings similar to themselves. Perhaps there are many kinds of creatures, each seeking a token of their own particular chemical make-up in the dim spectra of distant planetary systems. In a galaxy with over a hundred billion suns, like will attract like.

The Stirrings of Life on Earth

On the third planet from the Sun, a number of physical criteria were met that resulted in the spawning of life, some 4 billion years ago. The molecules that make up terrestrial organisms include hydrogen, carbon, oxygen, nitrogen, sulphur and phosphorus with minor amounts of metals like iron, sodium and magnesium thrown in for good measure. Biology tells us that inside every cell there is a wealth of simple, carbon-based molecules that have learned, through the propitious laws of nature, to become co-operative, self-replicating structures. The life-force can be rationalised by the natural propensity of organic molecules to form co-operative structures fuelled by a steady supply of energy.

As we have seen, the vast majority of the mass of the solar system was expended on the formation of the Sun and her attendant planets. As with any physical process, there were some left-overs, some residual matter that did not accrete into the great planetary bodies. This material consisted mostly of rocky and metallic substances, together with other, volatile materials, ranging in size from hundreds of kilometres to grains smaller than a humble bacterium. Most of this rocky material is still with us today in the form of *asteroids*, most of which are located between the orbits of Mars and Jupiter. Occasionally, small pieces of an asteroid pass very close to the orbit of the Earth and are subsequently captured by its gravity. Some of these meteorites have been accurately dated to 4.5 billion years ago, the epoch of the formation of the solar system.

The asteroids mark the boundary between the inner and outer solar system. Had Jupiter's gravity not been so influential, a small world a few hundred kilometres across would have accreted from the asteroidal material. Much farther out – half the distance to the nearest star, lies the Oort Cloud, made up of trillions of icy organic snowballs, the *comets*. Most comets lie in nearly circular orbits in the Oort Cloud, and so take many millions of years to complete one orbit of the Sun. Closer to home, astronomers have identified two other breeding grounds for comets. The first is located beyond the orbit of Neptune. Named after the Dutch planetary astronomer Gerard P. Kuiper (1905-1973), the *Kuiper belt* is believed to be the home of over a billion so-called *short period comets*, some of which grace our skies once every few decades. More recently, astronomers have identified yet another cometary home in the planetary part of the solar system. Located even closer than the Kuiper belt objects, the *Centaurs* are comet-like worlds located outside the orbit of Jupiter but within that of Uranus.

It must have been between 4.5 billion and 3.9 billion years ago that the Earth and the other planets in the Sun's family were continually bombarded by millions of meteoroids, asteroids and comets. The achingly beautiful vistas of the cratered moons of the giant gas planets, as well as those on our own Moon and on Mars, serve as poignant reminders of the terrible chaos that permeated the early solar system. Some of this doomsday debris may have been a few hundred kilometres across and would have caused immense damage to the early Earth. The solar system was a very dangerous place to be in those times of unsurpassed cataclysm, chaos and destructive indifference. But it wasn't all bad.

Comets also brought vast reserves of water and organic material to the surface of the Earth and may therefore have contributed to the rapid development of life. The meteoritic and cometary bombardment of our planet continued almost unremittingly for several hundred million years. We know that for the first 500 million years after the Earth's formation, our world was far too hot to support life as we know it. But gradually and imperceptibly, the incidence of celestial collisions abated. Then, suddenly, the Earth brought forth life in the form of simple, single cells, not unlike bacteria.

Palaeontologists have identified microfossil mats in rocks dating back 3.8 billion years. The cellular organisms found in these mats appear to be relatively advanced structurally, suggesting that the origin of life can be traced to perhaps 100 or 200 million years before these fossils were formed. Clearly, life must have arisen very quickly in geological terms. It may have taken only a few million years to get started. With the laws of physics everywhere the same, I wonder how many countless million times life has arisen spontaneously on other worlds in the cosmos?

There are a small number of scientists, some of them very influential, who believe that life did not originate on Earth, but was merely seeded here by comet or meteorite impacts. The idea that life came directly from the cosmos is surely an ancient one. Arguably the most vehement proponent of this so-called *panspermia hypothesis* is the English cosmologist, Fred Hoyle (b. 1915). Over the past twenty years, Hoyle has garnered information from a variety of sources suggesting that microbial life arrives continually from outer space. While it is known that the cosmos is replete with the substance of life, and that the Earth receives countless tonnes of material from space each year, there is no credible evidence that we are being continually visited by microbes from the cosmos. It is even more difficult to understand how these microbes could form in the frigid darkness of open space. What's more, the panspermia hypothesis does not answer the central question that biochemists want to ask. How did life *originate*?

Since life depends on the products of stellar alchemy, it follows that it can only be as common as the abundance of atoms that make up life forms. The early universe must have had far fewer life forms than it has today. This is because the heavier atoms that make up living things, like carbon and oxygen, take time to be generated in the nuclear furnaces of stellar interiors. Our Sun is a younger-generation star. Indeed, by the time it formed, there were enough heavier atoms around in the clouds of interstellar gas and dust to permit rocky planets with heavy gases, water and life to emerge. What would have happened if the Sun had spawned from first-generation stellar material? Without these heavy atoms, life would not have begun. The key ingredient for the formation of stars, planets and life is time. It takes time to form stars and these in turn need time to form the heavier elements that make up planets and life forms. Time is the thread that links all things.

How, precisely did life on Earth arise? The answer, in short, is that we don't know. Has life some special significance? Is it a phenomenon formulated from the familiar laws of physics and chemistry, or is there an intangible 'spirit' of life, forever beyond the ability of science to unravel? This is a question that has confounded generations of human beings, and only in the last few decades in the history of our species have we come close to answering some of the puzzles of the living world.

The hereditary substance for all advanced organisms is *DNA*, short for *deoxyribonucleic acid*. DNA is a polymer consisting of many individual, nitrogen-rich bases linked by simple sugar-phosphate bonds. A DNA molecule contains two long strands, which are complementary to each other. One strand can act as a *template* for the synthesis of a complementary strand of DNA enabling information within the molecule to be faithfully handed down from generation to generation. But how did this wonderfully intricate piece of molecular machinery, lying at the heart of every cell on Earth, come into being? To answer this, we must go back four billion years in our imaginary time-machine.

After the neonatal Earth had lost much of its primordial atmosphere of hydrogen and helium to space, it belched forth new gases, including carbon dioxide, methane and ammonia, as well as substantial quantities of water vapour. Slowly, our world cooled and the steamy, terrestrial atmosphere rained out the vast majority of its foggy vapour, forming the Earth's oceans. But the rains that fell on the primitive Earth were prodigious – not a mere 40 days and 40 nights, as the biblical Noah and his relatives allegedly witnessed, but rains that may have lasted for millennia. The consequent formation of vast expanses of liquid water on the surface of the primitive Earth was to prove crucial to the origin of life and to the later emergence of intelligence.

In this distant epoch, the young Sun was radiating ultraviolet light into our atmosphere – and at that time the skies of the Earth contained no ozone to shield us from these deadly rays. Some scientists have suggested that a haze of smoggy hydrocarbons may have peppered the upper atmosphere, acting as an effective shield against the deadly rays from our star. Then, as now, the Sun was an indifferent onlooker to the needs of the young planet Earth. And because there were no organisms, our world was as sterile as the Moon. Having just passed through its cratered adolescent phase, the Earth now boasted warm oceans of liquid water and a rich inventory of simple organic molecules.

Organic molecules are special. On Earth, they have a privileged status in the great orchestra of life. Organic molecules are made up of strings of carbon atoms, laced together with other atoms including nitrogen, oxygen, hydrogen, sulphur and phosphorus. Although we are pretty sure where the water and organic molecules came from, the question remains where and exactly how life on Earth sprang into being. There must have been many inland ponds experiencing a seasonal cycle of wetting and drying, ideal conditions for organic molecules to mingle and react to one another's company. Similarly, the ocean shores, with their diurnal tides, could have provided near-ideal conditions for the stuff of life to begin to assert itself.

But recent research is throwing new light on this great mystery. Oceanographers have discovered glamorous ecosystems of microbes and simple animals gorging on the energy-rich effluent gases and minerals belched from the Earth's interior at mid-oceanic ridges. Down there, in the abyss of the hydrosphere, the pressures are enormous, and save for the haunting light winnowing its way through the water from molten lava, the world is totally dark. What's more, the life that survives in these seemingly inhospitable places does not depend directly on the Sun. The most primitive bacteria so far identified by science have a love for broiling, anoxic broths of sulphur and methane. Give them oxygen and they shrivel up and die.

These wonderful revelations, together with the discovery of microbes living deep in the crust of our planet, are causing sea-changes in our understanding of the origin and evolution of life. Could life have first emerged deep within the Earth's interior? It's a provocative notion, given that almost all research carried out in prebiological chemistry (the study of the earliest molecular steps toward life) has been carried out on the assumption that life took its first steps in some forgotten pond or in the open oceans. The Earth's interior provides warmth, a means of concentrating biological molecules (*biomolecules*) and a shield from the ultraviolet rays that pummel the surface with a shower of death. In addition, it boasts a lavish inventory of minerals which are thought to be important in kick-starting the engine of life.

Regardless of the precise location of life's origin, it is thought that simple molecules were shuffled around in such a way that chemical reactions could take place between them, producing bigger, more complicated molecules. With time, certain biomolecules co-operated with one another forming primitive genetic systems, the first replicating molecules. Today, all life on Earth is based on four basic types of 'big' biological molecules – *proteins*, *lipids*, *carbohydrates* and *nucleic acids*. These, together with a few thousand smaller and simpler molecules, create biological animation from inanimation.

The Languages of Life

DNA is the vehicle of genetic inheritance. It is the pivotal nucleic acid of life, having the hefty responsibility of passing on all inherited traits. DNA is made up of a very large number of smaller molecules called deoxynucleotides. Hidden within the strands of DNA is the blueprint for an entire organism, be it a bacterium or a human being. Closely related to DNA is another polymer made up of a large number of ribonucleotides: *ribonucleic acid*, or *RNA*.

RNA acts as a messenger between DNA and protein. The instructions of DNA are imprinted or *transcribed* onto complementary RNA molecules. These in turn, through a process called *translation*, provide the formal instructions for the synthesis of proteins, the workhorses of the cell. Most proteins act as biological catalysts, or *enzymes*, speeding up chemical reactions within the cell. RNA can thus be considered as a molecular scribe, translating the language of the DNA into the molecular vocabulary of proteins.

The complex organisation of modern living cells evolved over hundreds of millions of years. For example, the flow of information from DNA to RNA to protein may not have always been the rule. Very early in the history of life, RNA evolved the ability to replicate itself and also, in the form of *ribozymes*, to carry out a variety of catalytic roles. Ribozymes are still in existence today inside our cells and provide us with an insight into the biochemistry of ancient life forms, an early paragraph in the history of life on Earth.

All good things must come to an end and so too did the reign of RNA. Natural selection produced a molecule closely related to RNA, which was slightly more stable under the prevailing geological conditions on the early Earth. By virtue of its greater reproductive fidelity and stability, DNA was found to be a better molecule for the purpose of transferring genetic information. Primitive living systems that were capable of synthesising DNA were able to replicate better and thus were selected over other life forms possessing the inferior RNA genes. In time, this selective pressure led to DNA completely superseding RNA as the molecule of heredity.

Meanwhile, nature was experimenting with proteins. These, like the nucleic acids, are large polymeric molecules, composed of simple amino-acid monomers. Proteins heralded a revolution in the history of life. Because proteins boast 20 different types of monomer, whereas RNA, like DNA, contains only four, they turned out to be far more efficient and diverse catalysts than ribozymes could ever be. Through evolution, the use of simple protein catalysts thus superseded the use of RNA as the biochemical workhorse of living things.

Over millions of years, primitive living systems organised their DNA into long, double-helical strands encoding discrete pieces of information along their length. The DNA was segmented into distinct *genes* containing all the information needed to synthesise the protein molecules required for optimal cellular function. In this unrelenting game of cosmochemistry, there must have been countless failures. Many replicating, semi-live molecular species must have perished as the moody forces of nature continued to select others that were better able to do all these things under the prevailing geological conditions. But in the end, nature's latent talent to spawn life won out. In this vast chemical cauldron that was the primitive Earth, mother nature came up with astonishingly complex, self-organised systems enclosed in a lipid membrane. This is the unit of life – the cell.

The Radiation of Life on Earth

During the first two billion years after the emergence of the first life forms, evolution came up with many different species of primitive cell. Despite their cellular diversity, these early organisms possessed a very similar architecture. Unlike the cells of our bodies, simple bacterial cells have relatively little internal organisation. Their bacterial *genome*, or packet of genes, is not separated into compartments but is instead anchored loosely to the cell membrane. All the complex biochemistry undertaken by bacteria takes place either in the membrane, or in the intracellular compartment known as the *cytosol*. There are no regions of the bacterial cell that are assigned to particular chores; it has no apparent division of labour.

Bacterial cells and all similarly designed cells are referred to as *prokaryotes.* In contrast, more complicated organisms such as single-celled yeasts and amoeba have a very different cellular make-up. Inside these cells, called *eukaryotes*, biologists can see many different compartments demarcated by an intracellular membrane. Such cells also tend to be much bigger, with many of the intracellular compartments comparable to bacterial cells in size. What are these subcellular compartments, and how could they be reconciled with the defined structure of the bacterial cell? Is there any continuity between bacterial cells and complex eukaryotic cells?

The first clue to the role of these tiny compartments came from studies in which they were removed from the cell so that they could be examined and manipulated. One example is the *mitochondrion,* an intracellular body having the dimensions of a typical bacterium. Biochemical analysis of isolated mitochondria revealed that they were capable of combining simple organic acids with molecular oxygen, releasing carbon dioxide as a result. In other words, mitochondria appeared to possess the ability to respire independently of any other cellular structure. Further painstaking work revealed that they even possess their very own DNA, encoding a number of enzymes that are required for the proper functioning of the intact mitochondrion.

It became clear that the mitochondrion was analogous to an organ of the body, carrying out a particular function. For this reason, biologists now refer to these well-defined structures within the cell as *organelles.* But it was the significance of the DNA that was to provide the clue to the origin of organelle structure. Through a method called *sequencing*, the precise arrangement of the bases that make up DNA was unravelled. Such methods, when applied to mitochondrial DNA, revealed that the pattern of bases was much more like bacterial DNA than any other form. This strongly suggested that mitochondria had a bacterial origin. The pieces in the evolutionary jigsaw puzzle were beginning to fit together. First, their size and shape are unmistakably like bacteria on both counts. Then, the DNA evidence demonstrates the clear homology they share with extant bacterial cells. Subsequent research revealed that other organelles were present within many animal and plant cells. *Chloroplasts*, for example, the organelles responsible for the harvesting of sunlight in the process of photosynthesis, were also found to possess DNA.

The key to the enterprise was first realised in 1970 by Lynn Margulis (b. 1938), a distinguished professor of biology at the University of Massachusetts. The main proposal of her hypothesis was that a species of large bacterium engulfed smaller bacterial species. But instead of eating its petite cousins, the bug lived in mutual harmony with them in a *symbiosis*. The formation of internalised, quasi-living organellar structures allowed for a remarkable division of labour to take place within these early eukaryotes, creating yet another dimension to the potential complexity of living things.

The Invention of a Cellular Sociability

For over 80% of the history of life on Earth, living things only came as single cells. But then suddenly, about 700 million years ago, the first primitive group of multi-cellular animals emerged – ancestors to today's polyps, hydras, medusas and jellyfish. Just how and why cells discovered and endorsed sociability is still something of a mystery, but as usual, there has been no shortage of ideas.

It has been suggested that multi-cellularity may have arisen from a failure of two cells to separate properly after nuclear division. Or, it may have been fuelled by a greater need to cooperate, perhaps in times of great hunger in the midst of some environmental catastrophe. Whatever the impetus for the aggregation of cells, they quickly established and exploited new ecosystems. Multi-cellular organisms could swim farther, explore more thoroughly and find new sources of food in times when their single-celled ancestors would perish.

With time, these cellular congregations stumbled on the idea of allocating particular chores to particular cells. For the first two billion years (or half) of the history of life on Earth, life consisted of nothing more than bacteria and other single-celled prokaryotes. While complex eukaryotes emerged from their prokaryotic ancestry some 2.0 billion years ago, it was not until 550 million years ago, or 86% of the way through life's history, that one witnesses the spectacular emergence of relatively advanced animal forms. This was a brief period of unsurpassed and unprecedented biological innovation, occurring over a period of just 5 million years. Biologists call it the *Cambrian explosion.*

The animal phyla formed during the Cambrian explosion provided the basis of nearly all the major body plans seen in the animal kingdom. Many palaeontologists have been puzzled by this extraordinary radiation of animal diversity, which occurred only half a billion years ago. Since the Cambrian explosion, evolution has merely built many variations on these half-billion year-old anatomical themes.

With the advent of multi-cellular organisms, cells became modified, leading to the emergence of organ specialisation. With the advent of multi-cellular organisms we also see the evolution of another vital attribute, destined to be crucial to the development of complex life forms: social behaviour. Because of increasing pressure to find sustainable energy sources, life forms exploited more and more niches of the planet. Driven by the indifferent vagaries of the Earth's climate, nature oversaw the emergence of animals – organisms capable of eating other energy-rich life forms, as well as plants – organisms preferring instead to harness the energy of sunlight, converting simple, abundant atmospheric gases and water into nourishing sugars. What a wonderful cosmic tapestry that brought forth the bacteria, protozoans, algae, molluscs, fish, amphibians, reptiles, plants, birds and mammals.

Whence all the splendour of the biological realm? Was it simply the product of the blind selection of favourable mutations, or the survival of the fittest, as Charles Darwin (1809-82) and his countless champions have openly and confidently declared for the past 150 years? Darwin's theory of evolution by natural selection, expounded in *On the Origin of Species* (1859), has been spectacularly successful in explaining a great deal. Most importantly, it demonstrates that the Earth and the riot of exotic life forms it supports have undergone ceaseless change, not over mere years or centuries, but over millions and billions of years. It clearly shows that we have evolved from a common ancestor, an ancient and complete organism that led to all future forms of microbe, beast and vegetable on planet Earth. It has shown that the accumulation of minor genetic changes over long periods of time can lead to impressive anatomical and behavioural differences between an organism and its distant ancestors. But the ever-active engine of evolution takes very long periods of time to produce something as humble as a bacterium. Furthermore, evolution relies on the fortuitous concurrence of myriad suitable physical and chemical conditions for the great symphony of life to be composed.

A century and a half on, can we place our hands on our hearts and announce that evolution by natural selection is the be-all and end-all of our outlook on biology? Is life, as the French biochemist Jacques Monod (1910-76) so depressingly insisted, just "chance caught on the wing?" Is the ancient covenant between Man and God in ruins or can we find solace in our contemporary views on the nature and origin of life on Earth? There are many scientists who are beginning to look at the phenomenon of life anew. In particular, many biologists and biochemists are beginning to sit up and take note of the astounding complexities of the subcellular world.

Where once we believed the cell was more or less a bag of protein enzymes with some compartments set adrift in a microscopic ocean of water molecules, we now appreciate the marvellous architectural details of the living cell. Inside every cell in your body lies a perfectly formed skeletal system composed of many hundreds of interlaced tubules and filaments. This beautifully constructed *cytoskeleton*, which bends and flexes to the cell's every movement, is enough to capture the admiration of even the most inventive engineer. These elaborate structures are capable of assembling of their own accord in a process known, appropriately enough, as *self-assembly*.

Self-assembly is a remarkable process. Genes make protein molecules, which fold into precisely the right shape to communicate and 'bond' with other proteins. Not only is self-assembly at work in the manifestation of the cytoskeleton, but it also underlies the mechanism by which a whole host of other complex cellular structures are constructed – from ribosomes that build protein molecules from RNA messages, to astonishingly complex motors, pumps and highly co-ordinated information-relaying systems. Are we missing the big picture? What is the source of all this complexity, this apparent 'knowledge' sewn into the fabric of the living cell?

Doubtless, an understanding of life begins with the precisely-tuned laws governing the fundamental particles of the cosmos. Through the seasons of geological time, matter and the laws of nature have allowed galaxies and stars, planets and life to emerge. We are in the wake of a new *paradigm shift*, which sees the universe's constituents merging and becoming more complex. Our mortal eyes have witnessed structure on the very largest and smallest scales – from the imposing walls of galaxies half a billion light years across to the awe-inspiring geometry of a six-sided snowflake. It seems that matter has a propensity to self-assemble by as yet poorly understood laws, allowing life and intelligence to flourish.

We seek new laws that reveal to us that we are 'expected', rather than "chance caught on the wing": this is the view of Stuart Kauffman, a philosopher turned biochemist at the Santa Fe Institute in California. Kauffman was deeply troubled by the posits of Darwin's theory and the way it is perceived by scientists and non-scientists alike. Drawing on philosophy, mathematics and evolutionary biology, Kauffman is in search of the laws of complexity. He believes that life occurs more or less spontaneously, once a critical level of molecular complexity develops in a system. Like the condensation of water from steam on a cold window, or the melting of the ice in the rivers and lakes of the Swiss Alps at the arrival of spring, the first true metabolism, Kauffman argues, came into being as a result of a molecular 'phase transition'. He believes that the 'complexification' of biological systems gives rise to *speciation*, and that natural selection is, by contrast, a secondary force, functioning to refine and elaborate what biological complexification achieves. In his book *At Home in the Universe*, Kauffman yearns to discover the laws that govern these activities.

Kauffman also believes that once these laws are known, it may be possible to create life from scratch – a bold prediction, given the utter failure of all those who have tried before him. Many biologists consider Kauffman's ideas to be overly complicated. Maybe so, but the basic principles which Kauffman sets forth are entirely reasonable and this has no doubt led to a growing armada of biologists who are casting doubts on the universality of putting Darwin's powerful ideas in the driving-seat of biological change.

My own reasons for doubting the supreme validity of Darwin's wonderful idea can be summarised as follows. Natural selection cannot and will never be a truly *universal* theory. The foundations upon which it is built are based on extreme improbability – the fortuitous concurrence of myriad physical and chemical conditions working blindly to bring about a living, evolving creature. Indeed, if all scientists believed in absolute Darwinism, no one would bother to look for life elsewhere in the universe. Life under Darwinism would be depressingly rare in the cosmos or even non-existent, except, of course, on this small world.

Darwinism can, however, explain the gradual changes that account for the variation in species. Thus natural selection can account for the differences between, say, two bacterial species, or two kinds of mammal. But it fails miserably in providing an explanation for the three notable phases in the history of life. We will define these phases as follows:
1. The emergence of the first cell.
2. The emergence of the first complex cells – the eukaryotes.
3. The emergence of complex multi-cellular organisms.
These are the three great transitions life has made in its 4 billion year legacy on Earth. What unites these transitions is their rapidity of emergence. Palaeontologists piecing together the early history of life on Earth are confident that it was kick-started very quickly, well within 100 million years and possibly within a few tens of millions of years. The same seems to be true of the eukaryotes. Cells very quickly figured out how to live one inside the other to form the first organelles. The Cambrian explosion also occurred rapidly, within 5-10 million years. These spectacular and, in the scale of geological time, abrupt advancements of life seem incompatible with the smooth, slow gradualism of natural selection. Something else must have fuelled these great transitions in the complexification of life.

The answer, no doubt, must have something to do with energy. There must have been a critical level of energy needed for a primitive metabolism to establish itself in the first place. This energy influx was responsible for the emergence of life. But of how it emerged, we can say next to nothing. The origin of life is still a profound mystery. But I believe we can say more about the two later phases of life.

The origin of eukaryotes coincides more or less with the emergence of the first photosynthetic cells a little over 2 billion years ago. Because the earliest forms of life emerged in a world devoid of oxygen, they had to rely on other gases to fuel their life-functions. But anaerobic metabolism isn't very good at extracting the chemical energy from fuel molecules. Cells capable of using oxygen could extract about 19 times more energy from a sugar molecule than their anaerobic counterparts. This, in my opinion, is no coincidence – the impetus for emergence of the eukaryotic cell may well have been provided by an increase in energy made available to it by aerobic metabolism. With more energy, a cell can build more complex molecular machines and scaffolds. It can apportion this energy amongst the various organelles making up the cell's structure. With more energy a cell can become more *complex.*

What can we say about the Cambrian explosion? As it turns out, scientists studying the early atmosphere of Earth believe that the oxygen levels reached about 15% (as opposed to 21% today) at the time of the Cambrian explosion. But physiologists have shown that complex animals need at least 15% atmospheric oxygen content to maintain their body functions. Could this oxygen threshold have fuelled the emergence of complex creatures such as those that exploded onto the world's stage half a billion years ago? More importantly, are we neglecting simple things like an organism's energy considerations in our assessment of how complexity arises in the natural world? Indeed, as we shall see in Chapter IV, oxygen metabolites have the ability to cause mutations which can fuel changes in complexity. Organisms can only become more complex if they have greater energy reserves, and also a means of changing genetically. Oxygen may well have provided both. The emergence of prokaryotes, followed by eukaryotes and multi-cellular creatures represent distinct *complexity levels.* I see natural selection governing changes *within* complexity levels, but having little to do with the 'complexification' events themselves.

Reflections

Most of the evidence we have concerning the nature of this vast universe paints the distinct picture of order and interrelationship. We live on a small, rocky world orbiting a typical yellow dwarf star in a 100 billion-strong community of other suns in the Milky Way Galaxy. As many as 60% of these stars may have planets. Why not life also? We are in search of a truly universal theory which makes life and mind an expected, rather than an astronomically improbable event. We are in search of a cosmological theory of biology. We have not yet found it, but there are reasons to think that we might soon.

We have now come to understand that we and the plethora of other organisms that inhabit this lovely little world are the products of 12 billion years of cosmic evolution. We, the products of an amorphous stellar gas cloud that once boasted nothing more complex than a tenuous conglomerate of atoms. Life is the most poignant and eloquent expression of universal law. It is here that we must begin to express our reverence for the universe and in time appreciate our true perspectives as we transcend our cosmic puberty.

Chapter II

The Manifestation of the Cosmos that is Man

Know then thyself, presume not God to scan;
The proper study of mankind is Man.
Placed on this isthmus of a little state,
A being darkly wise and rudely great
Excerpt from 'Know Thyself'
Alexander Pope (1688-1744)

Man is a physically unremarkable animal with ordinary powers of agility, speed and strength. Were it not for our brain, an extraterrestrial visitor surveying the worlds in our solar system might consider our species rather uninteresting. Perhaps some of these visitors would think our form strange and unfamiliar. For one thing, we are bipedal apes, a feature that is rare in the mammalian world. *Homo sapiens*, like all other life forms on the planet Earth, is the product of billions of years of natural selection and complexification. Until very recently, the forces of nature responsible for driving the evolution of the human species were identical to those governing the emergence of every other organism, from bacteria to blue whales. Yet in a remarkably short time, nature manifested an entirely different sort of creature. An ape that could reason and imagine, build tools and make abstract correlations of the world through art, music, mathematics and literature – a creature endowed with such strange attributes as consciousness and morality. Whence a creature like *Homo sapiens*?

In order to understand how we have come to acquire these extraordinary intellectual skills, we must take ourselves back in time to an epoch before the age of mammals, back to the end of the Cretaceous period about 65 million years ago. Here we shall begin our story of the ascent of the human species.

Setting the Scene

65 million years ago the predominant forms of life on Earth were the great reptiles: the flying pterosaurs occupying the skies, the ichthyosaurs, who reigned supreme in the seas and oceans of the Earth, and the dinosaurs who dominated the land. Together, these magnificent creatures occupied the pinnacle of the food chain in almost all environmental niches.

The dinosaurs were carnivores, herbivores and omnivores and they ruled the Earth for over 140 million years. Such successful creatures would have seemed invincible at their peak at the end of the Cretaceous Period, when an environmental catastrophe of gargantuan proportions wiped them off the face of the Earth utterly and forever. 65 million years ago these reptiles of the land, sea and air suffered a sudden and horrific mass extinction. In fact, over 70% of all plant and animal species were extirpated by the catastrophe. This holocaust has been dubbed the 'great dying'. What sort of global change could induce such a horrific death toll?

In 1979, the American palaeontologist Walter Alvarez (1911-88), while studying the sedimentation rates of metals in rock layers, discovered that the concentration of the rare earth element iridium was 30 times higher than normal in a thin sedimentary layer marking the end of the Cretaceous period and the start of the later Tertiary period. There were only two possibilities that could account for such an astonishing enrichment of iridium. Either a remarkably accelerated rate of deposition of this metal took place as a result of widespread geological activity, or it was deposited by an unusually iridium-rich source of extraterrestrial origin.

Although localised geological phenomena could account for a 30-fold enrichment of iridium, a local source could not be reconciled with subsequent discoveries, which showed that the iridium layer was present at various widely-separated locales all over the world. It is exceedingly unlikely that the Earth experienced a global geological change in these comparatively recent times. So where could the iridium have come from? A very important clue came from an unlikely source, the study of the chemistry of meteorites. Chemical analysis of many stony-iron meteorites revealed that some were highly enriched with iridium. These discoveries were to lead Alvarez and his team to propose that the most likely source of this iridium was extraterrestrial in origin. The collision of a 10 kilometre asteroid with the Earth, Alvarez argued, would inject sufficient iridium into the terrestrial atmosphere to account for the sudden deposition of the element at the end of the Cretaceous period, 65 million years ago.

The impact of such a collision would have been enormous. The energy would have been greater than the combined destructive power of all the nuclear weapons ever constructed. The explosion would have sent billions of tonnes of material high into the atmosphere, where air currents would quickly distribute the dust all over the Earth. This dust would have been concentrated enough to blot out the life-giving light from the Sun for several months or years. Our world would have experienced something akin to a nuclear winter – a long period of continual darkness and freezing cold temperatures. Plant life, so dependent on the Sun, would quickly die. Plant-dependent, herbivorous dinosaurs would also die out, and so on up the chain, eventually affecting the carnivores.

There have been suggestions that the 'great dying' was caused not by a single event, but instead by a series of closely spaced collisions. In this scenario a flurry of comets are drawn to the inner solar system as a result of gravitational perturbations caused by a passing star. Others have suggested that our Sun and its family of comets experience slight perturbations in gravitational tidal forces, as our solar system orbits the galactic centre. This activity might cause regular pulses of cometary approaches over geological time. Other scientists, such as the American physicist Richard Muller, have suggested that our Sun is not a single, isolated star, but is instead part of a binary system. Its companion, Muller believes, would be a brown dwarf star – a cool L-class star several dozen times more massive than Jupiter. This star, nicknamed Nemesis, is hypothesised to orbit the Sun about 1.5 light years away, allowing it to periodically plough through the Oort Cloud, sending a herd of comets to the inner solar system. Thus far, no Nemesis star has been found.

Several spectacular collisions over a period of a few hundred thousand years may have caused the dramatic demise of the dinosaurs. However, research in recent years has revealed, straddling the shore line of the Yucatan peninsula of Mexico, the likely impact site of a single gigantic asteroid. The Chixulub Crater has an estimated diameter of 180 kilometres. An impact in such a location would most likely have generated a huge tidal wave together with an enormous dust cloud. What's more, the object that created this impact site appears to have collided with the Earth at an angle, with the result that most of the energy from the collision, instead of dissipating in the Earth's crust, would have been unleashed into the atmosphere, inducing hideous changes to the Earth's climate.

Whatever the event, or events, that brought about the extinction of some 70% of the Earth's biota at the end of the Cretaceous period, it was almost certainly responsible for the emancipation of mammalian evolution. The most advanced species of mammal at that time would have resembled modern-day moles and shrews, but the numerous environmental niches opened up to them by the disappearance of their deadly adversaries would have caused an explosive radiation of these creatures, fuelled by the need to adapt to new habitats. The 'great dying' rang the death knell for most forms of life on Earth, but for mammals, it represented an exhilarating liberation, unsurpassed in their evolutionary history. The radiation of mammalian species led to all the bovine, canine, feline, cetacean and primate species found on the planet today. This serendipitous reshuffling was also responsible, at least in part, for the emergence of the genus *Homo*.

Everyone knows that we are related to monkeys and apes in some way. They share a very similar anatomy with human beings: the arms, the grasping and tool-using hands, as well as the familiar human-like gaze in their eyes. These physical similarities are also reflected in the language of genes. With the advent of molecular biology in the late twentieth century, we have become acutely aware of some startling facts. DNA sequencing has revealed that, on average, chimpanzees and humans are 99% similar at the genetic level – smaller than the genetic difference between a horse and a zebra. Palaeoanthropologists, using the mute fossil record of stone, have traced our ancestry back at least five million years. The fossil record is fragmentary and incomplete, but by studying extant non-human primate behaviour and biology we can recreate at least some of the events that shaped the evolution of our species.

The Mute Testaments to Human Prehistory

What are fossils? How can we interpret the origin and age of a fossilised organism? Imagine a scene, many millions of years ago, where a human-like ancestor may have lived and carried out her daily routine on the bank of a deep, sluggishly flowing river. Now severely arthritic, she makes her way to the water's edge as she has done many thousands of times in her long and productive life. After all, she has given birth to three fit and healthy males and has seen one of them grow to an age of self-sustenance. She slowly enters the water and stares curiously into its still reflections. There she sees the cloud-laced sky beneath her, and looking ever more closely, she notices a face in the waters. Startled, she retreats, but over-whelmed with curiosity and an unbridled urge to seek meaning in this strange image, she again returns to the restful river to search for the haunting image of an old female. Looking deeper and deeper into the waters, she moves her head from side to side and notices, to her aston-ishment, that the image moves with her. The cogwheels of reason turn in her mind as she further investigates the strange apparition in the river. Placing her hand over her head causes the water creature to do the same. In a moment of reflective rumination, she realises that she is look-ing at herself, her very own reflection. The graceful movements of her form consolidated in her a sense of self – the dawn of human consciousness.

Now let us imagine that, as this hominid bathes in the murky waters, she becomes exhausted, stumbles and drowns trying to recover from the fall. Her corpse, now fully saturated with water, both on her body and in her lungs, sinks to the tranquil river bed, where it slowly putrefies, leaving only the bones as a testament of her past existence. These bones are slowly covered by successive layers of sediment from the slow-flowing river. Over many thousands of years, the bones slowly exchange mate-rial with the surrounding silt and sand grains. After hundreds or thou-sands of years this exchange is completed. The bones, their structure, form and aspects of their history are said to have been *fossilised*, allow-ing them to be examined by intelligent minds hundreds of millennia after the events that marked the death of this once living individual.

The location of a fossil embedded in rock can be a valuable clue to its age. Geologists have known for many years that the chronicle of geological events can be traced by studying the manner in which successive layers of mineral deposits are built up. Many patterns of change in climate, the atmospheric composition and the plant and animal species inhabiting the Earth in past epochs can be deduced from a detailed examination of these successive layers, or *strata*. In addition, scientists have discovered a very simple way to date objects located within a particular stratum. This is achieved through a revolutionary technique called *isotope dating*.

Radioactivity - the Key to Ageing Fossils

Let's put our chemistry 'hat' on for a few moments. Imagine an atom, composed of electrons, protons and neutrons. In the early years of this century it was realised that the number and location of electrons orbiting the atomic nucleus largely dictate the chemical properties of an atom. Chemistry has very little to do with the nucleus itself.

One way to describe a chemical element is by calculating its *mass number*. This is the total number of protons and neutrons in the atomic nucleus. To maintain electroneutrality, atoms ordinarily possess the same number of positively and negatively charged particles, the protons and electrons. But no such rules exist for the number of neutrons that an atom can possess.

Over the aeons following the Big Bang, nuclear fusion reactions in stars created several forms of the same element, possessing an identical number of electrons but a different number of neutrons in their nuclei. Atoms with different numbers of neutrons in their nucleus but with the same number of electrons are called *isotopes*. Different isotopes of the same element share identical chemistry but display different degrees of stability. Unstable isotopes decay to simpler, more stable ones by ejecting subatomic particles from their nuclei. For example, some atoms spit out helium nuclei composed of two protons and two neutrons – the *alpha particles* – while others eject electrons, known in this context as *beta particles*. Different isotopic species have different characteristic rates of decay. Some elements have isotopes whose constituent atoms decay in a fraction of a second, while others possess atoms which decay over a very long period of time – over hundreds or even thousands of years.

As the chronicle of cosmic events has revealed, carbon atoms play a central role in terrestrial life forms. Carbon too has a number of isotopes. For example, the most stable and therefore the most abundant isotope of carbon has an atomic weight of 12. That is, it possesses six protons and neutrons in its nucleus and is referred to as carbon-12 (C-12). However, nature has also produced a carbon isotope with six protons and eight neutrons, carbon-14 (C-14).

A method of radiocarbon dating has been devised to assign an age to a fossilised sample. The principle is based on the decay of C-14 atoms to nitrogen by the emission of a beta particle. Because C-14's *half-life* – the time taken for half of its atoms to decay – is 5,730 years, this isotope must be continually produced or clearly it would already have disappeared. The location of its renewal is the upper atmosphere, where cosmic ray nuclei from the Sun and beyond bombard atoms of nitrogen and synthesise C-14 in the process. The newly formed C-14 re-enters biological systems via the carbon cycle. It can combine with the oxygen from the air to form C-14-labelled carbon dioxide. This is taken up by plants to make sugars, which in turn are consumed by animals including humans. As a result, animal tissue is slightly radioactive and when death occurs the C-14 has no way to get to the upper atmosphere. The C-14 levels within the dead organism begin to decline. By comparing the activity of a sample of fossilised tissue, such as a bone, with that of a living sample of the same tissue, an age can be computed from the ratio of the two values obtained. These isotopic dating techniques provide the most accessible way of ascertaining the age of a fossilised organism.

Biology Lights the Way!

The genetic blueprints given by the DNA sequences of a human and a chimpanzee are 99% similar. This fact alone is an awe-inspiring revelation. Inside every chimpanzee there may be the genetic potential to create a human being. Using dating methods based on the rate of mutation of vital proteins in a large variety of plant and animal species, it is now possible to estimate how long ago the human species diverged from the other primates. In the late 1960s, two American molecular biologists, Vincent Sarich and Allan Wilson, carried out detailed immunological comparisons of blood proteins from humans and African apes. To the astonishment of the anthropological community, their results suggested that the genus *Homo* diverged from non-hominid primates as little as 5 million years ago. This was in stark contrast to the contemporary estimates of 15 to 30 million years for such a divergence.

Further work by other molecular biologists has since pushed this date back to 7 million years ago. Nonetheless, the fact that our species, with its technological prowess, morality and outward-bound spirit, diverged from all other non-human primates only 7 million years ago is a testament to the extraordinary powers wielded by the forces of evolution.

What fuelled the birth of the genus *Homo*? By collating evidence from a variety of sources, including geological data and the fossil record, it has been possible to deduce something about the climate of the time.

Seven million years ago the African continent was lush with the verdancy of densely populated trees. Most of the extant species of ape were arboreal foragers, gracefully going about their business, relishing the succulent fruits that the forests bore them. At this time dramatic climatic changes began to assert themselves on the terrain of eastern Africa that were to irrevocably alter the behavioural patterns of at least one species of ape. The densely forested habitats were dramatically transformed into savannah. The density of trees in the region was reduced substantially, creating open country, which provided no protection from the heat of the afternoon sun or cover from the voracious predators that migrated to the region with the opportunistic herbivorous herds of wildebeest and Thomson's gazelle. Necessity is the mother of invention. In order to survive in this relatively hostile environment, our ape-like ancestors had to adapt.

What sort of attributes does an omnivorous, large-brained mammal need to evolve in order to survive in this strikingly different open-country environment? For one thing, the animal would need to protect itself more effectively from the unrelenting heat of the African sun which would have a high altitude for much of the day. Also, because of the longer distances between potential sources of food, the ape-like creatures would ideally need to carry their edibles in an efficient manner, perhaps by freeing both hands. Such pressures, imposed by the need to survive, prompted the incautious mechanisms of evolution to select individuals with the ability to stand upright, to become bipedal apes. This was to confer enormous advantages on these species as we shall see.

Traditionally, our cultural indoctrination has made us believe that the transition to fully modern, bipedal locomotion was gradual. The literature of the last 150 years abounds with cartoon figures depicting a sequence of figures, beginning with ancient human ancestors in distinctly stooped postures and ending with the anatomical omega of the form, the fully upright modern human. These very popular views of our ancestors have prevailed deep inside our minds for too long. At last we are casting aside these decrepit ideas about our evolutionary heritage. Recent evidence from a variety of sources, which includes careful reconstruction of fossil skeletal material as well as ongoing anatomical and physiological research, has revealed that it is highly unlikely that our ancestors experienced a gradual development of bipedal stature. It happened quite suddenly.

Contrary to our entrenched preconceptions, it is energetically and anatomically costly for apes to remain in a semi-bipedal or arched posture for long periods of time. Many palaeoanthropologists now believe that the sudden changes in the eastern African environment led to the selection of mutations in one or a number of ape genes that define the posture of the spinal chord and surrounding vertebrae. The result was a rapid transition to a fully upright ape with distinctly human locomotive patterns. This evolutionary development allowed the creature to use both its hands enabling it to manipulate its surroundings more effectively. Through the freeing of our hands, the human species has learned to communicate across the millennia. Our hands are the key to all our past and present technology. It may well be true, to a fair approximation, that what we do with our hands will determine the destiny of our species. We shall return to the human hand when we discuss the evolution of technology and the human brain.

Apart from the ability to forage for longer and to carry food over greater distances, what other possible advantages could be conferred upon an organism to help it survive in its new environment? It turns out that it's a question of cooling. Laboratory experiments have provided clear evidence that the adoption of an upright posture enabled these bipedal apes to better withstand the heat of the equatorial African sun. Quadrupedalism would mean that almost the entire surface of a creature's back was exposed to the searing heat of the Sun, whereas with bipedalism the overall body exposure is considerably less.

Based on these findings, some anatomists have suggested that bipedalism provided an efficient cooling mechanism for the animal, allowing for the accelerated evolution of the size and complexity of the human brain. Bipedalism may have even given these creatures a stalking advantage over other predators. Maintaining an upright posture bestowed upon these early hominids the singular ability to peer over long grass, to spot potential prey. This activity was increasingly important to these primates who had left their arboreal way of life far behind to seek a livelihood on terra firma. Indeed, the ability to scrutinise prey well in advance called for a commensurate increase in the complexity of hunting strategies, which in turn may have enhanced mental powers.

Let us now begin to build up a picture of the 'personalities' behind the fragmented fossil record of human origins. We must constantly bear in mind that our knowledge of the fossil record is incomplete, but we have unearthed enough pieces in the puzzle to allow us to build a reasonable picture of the emergence of our species. The earliest upright creatures that we know about are the *australopithecines,* literally, 'southern apes', which inhabited eastern Africa between 5 and 1 million years ago.

There were many species of australopithecine. All were more or less characterised by their bipedal anatomy and their small brain size, roughly equivalent to that of a modern chimpanzee. Their skulls were generally long and flattened with flaring nostrils. They also had disproportionately long arms resembling those of non-human primates. The anatomy of a variety of australopithecine species suggests that they maintained the ability to climb and move around in trees with the agility of modern great apes. As a result, it is reasonable to conclude that the change from a truly arboreal way of life to that of an upright land-dwelling existence may have occurred not long before this time. Estimates for this abrupt, but all-important change in behaviour are understandably various, and are put at between 7.5 and 5 million years ago.

The alarming number of newly discovered australopithecine fossils obtained by intensive site excavations in the first half of this century, impelled scientists in the early 1950s to subdivide the varieties of austral-opithecine into two large groups based on differences in jaw and teeth anatomy. These were the *gracile* australopithecines with their smaller jaw bones and less eroded teeth, and the *robust* species, characterised by their powerful, muscular jaws and severely worn dentition. It is clear that both groups relied heavily on plants in their diets, particularly tougher foods such as nuts and other hard fruits.

Two notable species have become instrumental in characterising the early bipedal hominids: *Australopithecus robustus* (*A. robustus*) and *Australopithecus africanus* (*A. africanus*). Comparison of these spe-cies' skulls reveals striking differences in their form. For example, *A. robustus* had a prominent crest spanning the top of its skull and had very large, protruding cheekbones with thick jaws. The *A. africanus* skull, on the other hand, had a smooth cranium with far less protrusion in the jaws. If we were to imagine these creatures going about their daily activities, we could not help but notice striking differences in their appearances. To the human eye at least, the robust variety would have been terrifying, their countenance unpleasant and disturbing. On the other hand, if we were to come across a member of the gracile variety, with its gentle facial features and hauntingly familiar behaviour, we would arguably concede that the descendants of these creatures led to the emergence of the genus *Homo*.

The robust australopithecines are now believed to have entered an evolutionary dead-end. Studies indicate that the robust australopithecines became specialists in their food requirements. Their extremely muscular jaw anatomy suggests that they specialised in eating tough, fibrous foods, unlike the gracile species, which consumed a greater variety of foods. As a result, when, as evidence suggests, significant climatic shifts occurred in central and eastern Africa about 2.5 million years ago, *A. robustus* and other robust species may have been driven to extinction by a dwindling of their food supplies.

Palaeoanthropologists think that *A. aferensis*, a species closely re-lated to *A. africanus*, became the direct ancestors of humanity. By far the most acclaimed example of *A. aferensis* is the skeleton of a female gracile australopithecine named Lucy, discovered by Donald Johanson and Maurice Taieb at Hadar, Ethiopia in 1974. The discovery of this largely complete skeleton caused shock waves in the anthropological community. Here was a hominid-like biped, standing over a metre tall, who lived 3 million years ago.

Skeletal analysis has revealed that, although this creature was well-adapted for bipedal locomotion, she retained relatively long arms and short hind-limbs. What's more, she possessed large, curved ape-like phalanges on her feet, suggesting that, although her main mode of loco-motion was bipedal, she still maintained a number of anatomical features of tree-dwelling apes. This could mean that Lucy represented a species of *Australopithecus* that was not fully ground-dwelling, but spent a sub-stantial percentage of its time foraging in trees in the spirit of its fully arboreal ancestors. Lucy did not forget her arboreal ancestry.

Despite the human-like features of the bipedal australopithecines, their brain sizes were very similar to those of modern great apes, between 400 and 530 millilitres; modern humans have a cranial capacity of 1,200 to 1,400 millilitres. Yet we have reason to believe that *A. aferensis,* at least, may have taken one step closer to becoming human. The earliest example of stone tools dates to about 2.5 million years ago, and includes crude scrapers, choppers and flakes. These were almost certainly used by an australopithecine species. Although we have known for some time that non-human primates such as chimpanzees use some very simple tools to facilitate their acquisition of food, there has never been a documented case of an extant non-human primate manufacturing stone tools from less refined raw materials. This suggests that australopithecines had a more advanced mental picture of the world than modern non-human primates.

I cannot help but wonder how these creatures lived and behaved. It is conceivable that they adopted lifestyles similar to modern baboons but with more emphasis on bipedalism. In the cool of the morning, perhaps a small group of australopithecines would disperse to forage on the ground. When a lucky individual spotted some lush fruit growing on trees, it was equally adept at exploiting this arboreal food source. Of course, formid-able climbing ability would also stand them in good stead in times of danger, such as in an encounter with a neighbouring group of australopithecines or a hunting pack of hyenas. Returning home in the late afternoon, they might contemplate the days ahead and consciously select simple stone tools to aid them in their tasks. I suspect that the later species of *Australopithecus*, at least, had a rudimentary awareness of themselves, a form of self-consciousness that was subsequently nur-tured and enhanced in all descendants of the genus *Homo*.

Recent fossil finds in eastern Africa have pushed back the age of bipedalism to at least 4.5 million years, with the discovery of a significantly older partial skeleton, now known as *Australopithecus ramidus*. It is widely believed that bipedalism will be revealed in still older fossils, yet to be unearthed. Many palaeoanthropologists now think that bipedalism could have marked the anatomical dividing-line between hominids and the apes, which diverged some 7.0 million years ago. This leads us to conclude that since stone tool use was introduced into hominid society some 5 million years after the anatomical adaptation to bipedalism, the latter had little to do with the development of stone technology. Put another way, the evolutionary driving force behind the emergence of stone tool use cannot be reconciled with the adaptation to an upright posture. Yet, as we have seen, the change to a bipedal form of locomotion freed the hands for other uses and we can only assume that the earliest hominids needed another vital ingredient to take the next big step in becoming human – the enlargement of the brain. This is a unique feature of the genus *Homo*.

The Giant Leap: the Birth of the Genus *Homo*

By about 2.4 million years ago, the fossil record displays evidence for the emergence of a different sort of creature from the ape-like australopithecines. For one thing, the brain size of these creatures was almost twice that of any australopithecine species so far excavated. Clocking in with an average brain volume of 700ml, these hominids were the first to fashion stone tools and therefore represent the advent of the first truly recognisable human ancestors. Many incomplete fossil skeletons displaying the characteristics of this new hominid have been unearthed in southern and eastern Africa. Such were the anatomical changes evident in these fossils that in 1964 the palaeoanthropologists Louis Leakey, Phillip Tobias and John Napier named the specimen *Homo habilis (H. habilis)*, which literally means 'nimble man'.

Scientists have gathered evidence suggesting that several subspecies of *H. habilis* existed over a period of a million years. Most notable are the large- and small-brained *H. habilis* specimens excavated from Koobi Fora in Kenya. The larger species possesses a large brain and a broad, flat face with big teeth and jaws. In contrast, the smaller species, although more recent, is smaller brained, with a more protruding face and more human-like jaws and teeth.

Homo habilis is a worthy appellation for this hominid, representing as it does a giant leap forward in terms of brain size and behavioural complexity. Small but deeply significant anatomical changes in the hands of this creature produced opposable thumbs, conferring a greater degree of dexterity in the fashioning of stone tools. In fact, the anatomical make-up of its hands is essentially identical to that of modern humans. The striking increase in brain size must also have allowed this creature to manipulate its environment far more effectively than any australopithecine. We now know that *H. habilis* was a contemporary of at least three species of *Australopithecus*. Competition for food and water reserves must have been fierce, but the increased computational power bequeathed to our habiline ancestors gave them the competitive edge they needed. But exactly how human was *H. habilis*? In particular, was it conscious of itself and the world around it, and did it have some form of spoken language?

Neuroanatomists who have studied the modern human brain in great detail have mapped out regions of the cerebral cortex which seem to be important in dictating the language instinct. They have located a slightly raised lump in the left temple of the human brain that appears to be essential for the production of spoken human language. This is called *Broca's area*, after the outstanding anthropologist and physician Paul Broca (1824-80) who first identified it. Because this language centre undulates over the furrowed surface of the brain, it leaves a distinctive impression on the inside of the cranium, which allows it to be identified in fossilised skulls. This is achieved by making plaster casts of the inside of partial or intact hominid crania. Using this method, Ralph Holloway at Columbia University has shown that a well-preserved *H. habilis* skull possesses a raised locus inside its cranium corresponding to Broca's area. This language centre is conspicuously absent in all australopithecine skulls examined to date.

There is another way to deduce something of the linguistic abilities of our hominid ancestors. This has arisen from a study of the position of the larynx, or voice box, in the throat of fossilised specimens. In the chimpanzee, the larynx is located high in the throat, restricting the range of sounds that can be produced to a series of grunts and squeaks. In the human, however, it is located significantly lower in the vocal tract, which increases the repertoire of noises, allowing humans to express a far richer variety of sounds and thereby making complex language possible. In fact, studies of the developmental anatomy of human children have shown that at birth the larynx is located high up in the throat, but slowly migrates down the vocal tract as the child ages. This correlates very well with the increasing linguistic abilities of children as they grow older.

Anatomical studies of this kind have not been possible with *H. habilis*, as no complete fossil skeleton has yet been excavated. We can only deduce that habiline man had a capacity for some form of spoken language. This conclusion is supported by the forms of the stone tools excavated at *H. habilis* sites, which clearly demonstrate that the creatures were predominantly right-handed. In most modern humans, the left side of the brain is larger than the right, making approximately 85% of individuals right-handed. This *brain lateralisation* coincided with the emergence of the earliest tool-makers. Moreover, the neural machinery required for spoken language is located in the left hemisphere of the brain, suggesting that the capacities for tool-making and for human speech co-emerged in evolution.

There can be no question that *H. habilis* was to some extent con-scious and capable of foresight. It must have had a vivid mental picture of the tool it required, before creating it. To have had a premeditated mental picture of an object that had not yet been created plainly implies that these hominids had a sense of purpose, and hence a significant de-gree of self-consciousness. The deliberate fashioning of stone tools from raw materials is clearly an extension of the self, the inner 'I'. Conscious-ness has been with the human species for millions of years.

Build a Technology That Lasts a Million Years!
By about 1.8 million years ago, there is evidence in the fossil record for the emergence of another hominid species in Africa, *Homo erectus*. These hominids represented yet another great leap towards humanity. With an average brain size of 900ml, they were significantly larger than any specimen of *H. habilis*.

H. erectus may have been an unusually tall hominid. The skeleton of a *H. erectus* boy, estimated to be about nine or ten years old, was found at Nariokotome in Kenya and shows that he was already 1.7 metres in height and heavily built. The brow ridges in *H. erectus* were prominent, but the face was considerably flatter than in *H. habilis*. Based on the aforementioned anatomical criteria, *H. erectus* also had the capacity for spoken language.

We see the emergence of many technological innovations coinciding with the appearance of this hominid. The bifacial hand-axe, the cleaver and the use of fire are three notable manifestations of the mind of *H. erectus*. Such was the effectiveness of their weaponry that their tool technology, and the trade that resulted in its widespread use, endured for a million years. We also witness the emergence of the enterprising spirit of humanity in *H. erectus*. It was the first hominid to leave the African continent, migrating into Asia and Europe as early as one million years ago. Specimens of *H. erectus* have been excavated in locations as far apart as Beijing and France.

H. erectus was clearly a resourceful animal. There is evidence that these hominids made tools from materials that were locally available, such as wood, flint, lava and chert. They were omnivorous creatures, who adopted a hunter-gatherer lifestyle much like later generations of nomadic humans. The profound changes in social complexity of the *H. erectus* community over that of *H. habilis* suggests that they re-quired a more complex system of spoken language. The organisation of hunting and foraging strategies must have required more complex intel-lectual skills. The significance of *H. erectus'* use of fire is the subject of debate amongst some palaeoanthropologists today. Ash excavated from a *H. erectus* site at Zhoukoudian near Beijing strongly suggests that our million-year-old ancestors had tamed fire. They may have used fire to protect themselves from predators, or they may have used it to cook food. If they used fire purposefully, then it is reasonable to conclude that they had a more complex social structure than any earlier hominid species – a structure that may have required them to share food more efficiently. Sometimes I wonder what these people thought or spoke about, as they huddled to keep warm around their ancient camp-fires. Perhaps they wondered about their origins and the significance of their being. We shall probably never know.

The Enigmatic Neanderthals

By around 400,000 years ago, there seems to have been enough hominid diversity to allow further evolutionary developments to take place. This can be attributed, no doubt, to the success of *H. erectus* as an adaptable and migratory species which exploited many different terrains and climates of the African and Eurasian continents. Although some fragmentary fossil evidence suggests that a variety of other hominid species diverged from the nomadic *H. erectus*, there seems to be no consistent change in the fossil records until about 200,000 years ago, when a rather elusive hominid came to dominate the pinnacle of the food chain in Europe and western Asia. These were the infamous *Neanderthal* peoples.

Over the past century archaeologists have unearthed many specimens of Neanderthal hominids at locations as far apart as northern Iraq and southern Scandinavia. These late people were highly evolved humans but were probably not our direct ancestors. Nevertheless, archaeologists class them as an early species of *Homo sapiens*, that is, *Homo sapiens neanderthalensis*. Anatomically they were shorter and more muscularly built than modern humans. Their brains were actually larger than ours, having a cranial capacity of between 1,200 and 1,750ml (as opposed to 1,200-1,400ml for modern humans), but were flatter and broader at the front than in modern humans.

The Neanderthals in many ways have borne the brunt of our ignoble sentiments towards our ancestors. The early literature of the twentieth century abounds with depictions of Neanderthals as ape-like, brutish caricatures brandishing clubs and harassing weaker members of their community. But times are changing. We now know that they were fully upright people, and although they had prominent brow ridges with large, flattened noses, they would not look much out of place in today's society.

The Neanderthals made important technological advances in the use of stone tools and weapons. They were probably the first hominid species to have a sense of religion and may even have upheld a belief in an afterlife. This evidence has come form a variety of sources. For example, excavations have revealed that Neanderthal skeletons were buried face-up and with hands crossed, as if deliberately tied. We also know that

Neanderthals cared for weaker members of their community. Perhaps the most cited example is to the 'Old Man' of La Chapelle-aux-Saints. This individual died at the ripe old age of forty, a notable achievement in prehistoric *H. sapiens* biology! Recent evidence suggests that this poor character was afflicted by a variety of debilitating maladies including arthritis of the hip, feet, skull, jaw and spinal column, as well as abscesses and rib fractures. These afflictions most likely took place over a long period of time and would have compromised his ability to function normally in Neanderthal society. Therefore, he must have been nurtured and cared for by members of his immediate community to have lived as long as he did.

The Evolution of *Homo Sapiens*

The Neanderthals were highly successful hunter-gatherers, surviving for more than 200,000 years. But quite suddenly, by about 34,000 years ago, the Neanderthals disappeared from the face of the Earth leaving no recognisable descendants. Why did a species of highly evolved humans with a soaring spirit and organised culture suddenly become extinct? We still do not know the precise answers, but we have unearthed some important clues to this puzzle. By about 100,000 years ago, we see the emergence of a new type of human being in the mute fossil record. These creatures had a more slender musculature than any Neanderthal, had developed weaponry and tool technology to a level of sophistication surpassing most Neanderthal inventions, and apparently were the first to express abstract ideas in the form of art and sculpture. They were the early moderns – the direct ancestors of today's human species.

The early-modern human beings apparently evolved in Africa more than a 100,000 years ago. By about 40,000 years ago, they had migrated into the lands of Eurasia, traditionally occupied by the Neanderthal peoples. It is easy to get carried away with our imaginations here. Anthropologists have supposed quite naturally that a competition ensued between these two distinct species of human. There was no contest from the outset; the technologically superior moderns forcing the Neanderthals to live in ever more isolated regions of Europe until, by about 34,000 years ago, they simply became so marginalised that extinction was inevitable.

There is some evidence to suggest that this is what took place. But is it not possible that the Neanderthals interbred with the moderns? The evidence would suggest not. The customs and language of the two human peoples would have prevented any long-term relationship to be forged between them, putting into doubt the contention that they simply interbred and became 'fused' as it were. There are important anatomical differences in the location of the adult larynx between modern humans and Neanderthals which supports these ideas. Curiously, the Neanderthal voice box was located farther up the vocal tract than it is in modern humans and even in *H. erectus*. This may seem like a contradiction in terms, since in all other respects, Neanderthals were more anatomically akin to modern humans than to *H. erectus*.

The evidence would suggest that the Neanderthals were capable of engaging in a very rudimentary form of language, perhaps similar to that of a modern human child. Plainly, the differing capacities for complex spoken language between the two peoples may have prevented any interbreeding between them. The Neanderthals, it seems, relied more on their physical strength and athleticism than did the early moderns, who probably made more use of cognitive inventiveness to survive in the Europe of 34,000 years ago. Modern human beings clearly had a greater technological capacity than the Neanderthals, and it was this advantage in natural inventiveness that tipped the balance in our favour. The moderns survived to build the world we see today. From this species came the art, science, technology and soaring intellect that we recognise as quintessentially human.

The Story in the Genes

Recent molecular biological evidence has strengthened the notion that the Neanderthal peoples were fundamentally different from modern humans. Because Neanderthal remains are comparatively recent in geological terms, molecular biologists reasoned that it might still be possible to extract DNA from Neanderthal bones and compare it with modern human DNA. In animals, including humans, DNA comes in two basic forms, genomic and mitochondrial. Genomic DNA is very large, linear and thread-like in texture, while mitochondrial DNA is small, compact and circular. In practice, mitochondrial DNA is far more stable than the genomic variety, and although it is more difficult to extract for analysis, it yields potentially valuable clues to relationships between species.

The extraction of ancient DNA from fossilised biological specimens is fraught with difficulty, even at the best of times. How could one possibly go about analysing ancient Neanderthal DNA? We would have to make sure that there is no contamination of the samples with modern DNA. Added to that is the problem of the relative paucity of intact segments of DNA long enough to be unambiguously identified. Fortunately, there is a way to make the most of the available DNA from even a very ancient sample. What if we could somehow amplify the useful segments of DNA in a way that would allow us to characterise them easily?

In the 1970s, an American molecular biologist, Kary Mullis (b. 1944), was studying small bits of DNA in his laboratory. One of the problems he faced was that the particular DNA molecules under study were of low relative abundance in his samples. Such was his need to surmount the problem that Mullis dreamt up the idea of the *polymerase chain reaction*, or *PCR*. The idea of PCR is ingeniously simple, so simple in fact that it is all too easy to dismiss the discovery as being inevitable. Deep inside our cells, in the nucleus, there are enzymes that bide their time in association with DNA. One of these enzymes is called DNA polymerase. Its job is to take the individual nucleotide bits that constitute DNA and add them in sequence so that a new strand can be produced. In the last chapter we took a look at the DNA molecule. One strand and its constituent nucleotide bits bond with a complementary DNA strand. When DNA needs to replicate, one strand unwinds and acts as a template for the synthesis of a complementary strand. In this way the perpetuation of the species is ensured, at least at the molecular level.

Mullis had the presence of mind to wonder what would happen if he isolated some DNA polymerase from cells and used it to replicate the DNA of interest to him. Such was his motivation for surmounting the problem that he eventually hit on the solution – by using enzymes to amplify his DNA millions of times. He added the individual deoxynucleotide bits to his DNA sample, threw in some short stretches of DNA to 'prime' the reaction, and finally added DNA polymerase. He then subjected his samples to sequential rounds of heating and cooling, allowing the newly formed strands of DNA to separate from the template strands. By repeating this simple procedure several times, Mullis was able to amplify his DNA sample millions of times, allowing him to study it at his leisure. PCR has found almost routine use in all respectable molecular biology laboratories the world over.

Using PCR, scientists amplified a sample containing Neanderthal mitochondrial DNA. The results showed that there were significant differences between Neanderthal and modern human DNA. For example, unlike modern humans, whose mitochondrial DNA sequence varies by as much as eight individual bits, the Neanderthal samples fluctuated by as much as 27 bits. These results, together with a comparison with chimpanzee DNA, suggest that Neanderthals represented a species about midway between humans and chimpanzees. The greater-than-expected differences between the Neanderthal mitochondrial DNA sequence and our own also suggest that Neanderthals may have appeared as early as 690,000 years ago, and that they too originated on the African continent. It is very likely, therefore, that the Neanderthal peoples enjoyed a period of roughly half a million years of independent evolution before anatomically modern humans came on the scene. This is certainly time enough, in my books at least, to account for the profound cultural and physical differences between us and the Neanderthals. It also goes a long way toward explaining why Neanderthals didn't interbreed with modern human populations, and thus, why those people died out without leaving any recognisable descendants.

Brains, Glorious Brains!

The biological evolution of the genus *Homo* is truly remarkable. In just 3 million years our cranial capacity trebled. Surely this sudden and awesome enlargement of our brains is what sets us apart from the rest of the animal kingdom. Such a rapid evolutionary change in our brain size calls for an extraordinary explanation. We must look first to our genes for a rational interpretation of this event. As shown by the molecular biological evidence, our genes are, on average, more than 99% similar to those of our closest living relative, the chimpanzee. But the behavioural and cognitive differences between humans and chimpanzees are so great that they have been classified into two separate families. This suggests that the rapid divergence of human ancestors over the last 7 million years or so has come about from relatively few mutational changes in those genes involved in controlling the levels of particular biomolecules, such as nerve growth factors and other neurochemicals. Put another way, it is not gross genetic differences that distinguish humans from non-human primates, but rather biological decisions that dictate how much of each protein or neurochemical will be made, where it is produced, and when it is released to perform its biochemical duty.

The human brain has undergone a phenomenal increase in the size of the cerebral cortex. It is the cerebral cortex, providing the capacity for intelligence, that distinguishes our brain from those of other animals. It is conceivable, therefore, that mutations in the relatively few genes that regulate cranial capacity were responsible for the sudden enlargement of the human brain. It is difficult to pinpoint a likely environmental trigger for this remarkable cranial enlargement in the genus *Homo*. Perhaps it was a change in our diet that led to this arguably fortuitous change. Many plants contain natural chemicals that can cause genes to mutate. It is well known that the act of cooking food can produce a number of mutagenic substances. These agents may have led to the pronounced cranial expansion in humans. We simply cannot tell for sure, but I suspect it has something to do with abrupt changes in our diet.

As an event, the rapid and profound increase in the size of the human brain created much new space for neural processing. Perhaps initially our ape-like brain could not find use for this increased cranial capacity that apparently had been created quite by accident. Over many millennia, existing brain tissue innervated this new frontier, slowly finding novel uses for the serendipitous expansion. In a sense, humans are endowed with a brand-new organ for which we have only recently begun to find uses. The human cerebral cortex is a wonderful gift from nature. Its very existence is a poignant symbol of the power and supreme elegance of evolution. It is as if nature, desperately seeking to understand itself, has found a means to do so within the electrical complexities of the human mind.

But the enormous cognitive difference between humans and other primate species is not reflected in the genes. Perhaps the most striking way to reveal the extraordinary similarity between human and non-human primates is to examine the pattern of chromosome pairs in a variety of closely related species. *Chromosomes* are thread-like structures that are composed of DNA and protein. They provide a way of efficiently packaging all our genes into a very small volume. Every cell in the human body, except the sex cells, contains 23 pairs of chromosomes, consisting of one copy from the mother and one from the father. Scientists can see these structures by staining the chromosome pairs with a fluorescent dye that attaches to their constituent DNA molecules. This dye, when irradiated with light of a particular wavelength, will cause the DNA to fluoresce, rendering the fine structure of the individual chromosomes more visible. In this way, it is possible to compare and contrast the entire chromosomal assembly, or *karyotype*, from a variety of primate species.

Primate karyotyping results show that all the great apes have 24 pairs of chromosomes – just one more pair than in humans. Eighteen of the 23 pairs of chromosomes are identical in four species of primate – humans, chimpanzees, gorillas, and the orangutan. Of the remaining pairs, four show only slight variations between species. The differences are most pronounced between humans and the orangutans, while those of humans and chimpanzees are, to all intents and purposes, identical. It is only in the human chromosome pair number 2 that we see a major divergence between other primates and us. Some biologists have noted that the size and structure of the ape chromosome pairs 12 and 13 each match a portion of human chromosome 2. Indeed, it is now believed that these ape chromosomes fused to form the larger chromosome 2 in humans. Could these comparatively small changes in chromosome structure have fuelled the birth of the genus *Homo* from our primate cousins? Until we find out exactly what genes are located on these chromosomes we cannot know for sure.

I was once told an amusing tale about how humans came to be the way we are. The human species, it was asserted, was the result of tinkering with the genes of more primitive primate species by an advanced extraterrestrial civilisation. We are, it was claimed, the progeny of genetic engineering, constantly watched by ETs, indifferent to our fate as they hideously assess the fruits of their scientific labours. Although the idea is far-fetched and very probably not true, we shall soon have the means to decide whether any such genetic manipulation has occurred. The *Human Genome Project* – an international effort by scientists to sequence the entire human genetic code – will be completed in the first decade of the twenty-first century. All the genes that characterise a human being will be made available for intense study. This will represent an astronomical leap forward in our understanding of our species. If any signs of genetic manipulation are present in human DNA it will soon be revealed. And as for tinkering – we shall be doing the job of the extraterrestrials!

Homo – The Sexually Mature Baby Chimpanzee

As we have seen previously, humans and chimpanzees are astonishingly close in genetic terms. How can all the unique cognitive and linguistic capacities of human beings be reconciled with a mere one per cent genetic difference from chimpanzees? As I mentioned earlier, the fact that we share so many genes with our nearest relatives tells us that the genus *Homo* emerged from primate ancestors by undergoing a number of crucial mutations in genes that influence the regulation of body patterning. Let's take a little time to build upon this argument.

Early genetic work carried out in the 1960s on simple animals such as fruit flies (*Drosophila*), showed that fly embryos have eight body segments, each of which is regulated by a single gene. This idea went more or less unnoticed until 1984, when the molecular biologists William McGuinnis and M.S. Levine, at the University of Basle in Switzerland, published a paper showing that these *hox genes* were present in animals as divergent as flatworms and humans. Their astounding discovery clearly demonstrated that the same basic genetic software is responsible for orchestrating the body plans of all complex creatures – creatures separated by half a billion years of evolutionary time.

Further work showed that different organisms have different numbers of copies of these hox genes. Worms have one cluster, lampreys and other primitive vertebrates have two clusters, while fish have four. One might expect more clusters for more advanced animals such as birds, mammals and reptiles, but that is not so. Indeed, humans and all other complex animals on Earth have no more than four clusters of hox genes. These discoveries raise the intriguing question of how the profound differences in the anatomy and behaviour of human beings were achieved without the need for a greater number of these developmental genes.

Part of the answer, it seems, may be a matter of timing. Altering the timing of the 'firing' of these genes causes new developmental patterns to emerge, leading to astounding new ways of achieving diversity without increasing the number of genes. This process, called *heterochrony* by developmental biologists, may provide us with important clues to our own origins. Take the case of baby chimpanzees. Their skulls – the flat face, the large rounded brain case, the position of the eye sockets – bear a distinctive resemblance to adult human skulls . Indeed, many developmental biologists now believe that a change of timing in the activation of hox genes gave rise to many novel species bearing strong resemblances to the juvenile forms of our ancestors. We may very well be sexually mature baby chimpanzees!

Immature Arrivals in the Outside World

The pronounced enlargement of the hominid brain relative to other primate species posed a considerable threat to mothers giving birth to live young. The infant's head could not pass easily through the pelvis. Because there is a limit to how wide the pelvis can be, survival of both mother and baby required human babies to be born in a very immature state relative to other primates. A human new-born is far more helpless than any neonatal chimpanzee. Some biologists have suggested that humans, being the size they are, should have gestation periods of nearer two years, rather than just nine months. The benefits of giving birth to immature, defenceless young must have outweighed the short-term disadvantages. But what could these benefits be?

The steady growth of the human brain after birth calls for an extended period of learning, both cultural and intellectual. The period of immaturity for the chimpanzee is approximately eight years, while that of a typical human being is sixteen years. Although there must have been a considerable liability in the investment in extended periods of immaturity in hominid society, this was more than compensated for by the extraordinarily rich repertoire of social and intellectual skills acquired during the period spent in suspended youth. Human beings need a much longer childhood to find their bearings in a world with so many cultural complexities.

Homo sapiens is an extremely variable species. Some anthropological authorities have chosen to recognise as many as 30 races, defined as regional populations that differ genetically but can readily interbreed. It is precisely this genetic diversity among human races that makes us what we are. Moreover, although humans have not changed anatomically for over 100,000 years, our high technology would make us seem like magicians to our distant ancestors. Humans were just as smart then as we are now, but our world and theirs are separated by an ocean of difference – a difference brought on by the flowering of human culture. Cultural evolution proceeds much faster than biological evolution. As the great Russian-American evolutionary biologist Theodosius Dobzhansky (1900-75) said, "Man adapts his environment to his genes more frequently and efficiently than his genes to his environments."

The Moral Ape

A human being is much more than an intellectually gifted chimpanzee. We are also moral creatures – a trait that is conspicuously absent in other members of the animal kingdom. There are, of course, many examples of altruistic, or 'selfless' behaviour in evidence across the animal kingdom. Altruism is exhibited by some insect societies. Bee colonies, for example, have numerous male workers which are not only sterile, but which also protect the hive and the queen from attack by other bees and insects. Additionally, the male worker possesses a sting that, once inflicted upon an intruder, causes the worker to die. Bee societies have developed altruism to a supreme degree. We see more familiar forms of altruism in a variety of mammals and birds. Take the Bending squirrel, which inhabits the mountainous regions of the American west. These squirrels are normally preyed upon by hawks and coyotes. If a squirrel senses a predator's approach it sounds a high-pitched alarm call which alerts other individuals, allowing them to scurry to cover. But the very act of sounding the alarm places the squirrel in far more danger than normal, had it remained silent. This is amply borne out by detailed observations.

What then can we make of these acts of apparent selflessness exhibited in the natural world? Neo-Darwinists believe that natural selection favours those anatomical, physiological and behavioural traits which maximise reproductive success. When a parent risks her life for her offspring, the altruistic act serves to increase the chances of the parental genes' propagation in the wider community. But altruistic behaviour is not only lent to direct offspring. It can also be extended to offspring of other close relatives, such as cousins.

The leading theory that attempts to explain altruistic behaviour is called *inclusive fitness*, a term first used by the English biologist W.D. Hamilton in the 1960s. According to Hamilton, many animals show the greatest degree of altruistic behaviour toward their offspring and progressively less for related members of their extended family. By protecting offspring closely related to its own, Hamilton argues, a parent can ensure that a greater percentage of its own gene library will be propagated into the future.

There are, however, some rare examples of another form of altruism, which extends to individuals totally unrelated to the 'kind' individual. Baboons, for example, and other closely-knit mammalian social groups display the phenomenon of *reciprocal altruism* – helping unrelated individuals in the hope that they will reciprocate the gesture at a future time. This sort of tit-for-tat altruism is exhibited only where many individuals have a chance to exchange aid.

But how does this theory fit with the high morality and ethics of the human species? Although I would concede that many aspects of basic human behaviour might exhibit a tit-for-tat altruism, there are also many examples of selfless and courageous acts in humans which cannot be explained in this way. For example, I am fully aware that, when visiting a strange city, my charity in giving a vagrant some loose change cannot and does not affect my long-term ability to survive. Indeed, it is highly improbable that I will ever encounter this particular individual again. What, then, is my motive for showing kindness? Furthermore, biology may have something to say about altruism towards other members of an animal's kind, but human beings are capable of extending compassion and sensitivity to other species. As a leading author of an undergraduate biology text put it, "true altruism never really occurs, except perhaps in humans."

And what about celibacy? What is the Darwinian motive in consciously resolving not to procreate for the sake of attaining a greater good? Most people would argue that celibacy is unnatural in the sense that most people like to form new families and have their own children. The act of celibacy is an example of the way in which human beings, empowered with the attribute of morality, can override the most fundamental driving force (according to Darwinism) in the biological universe – our libido. It is these and other paradoxes in the nature of human behaviour that raise serious doubts on the validity of Darwinian evolution as a fully comprehensive theory of human origins. In short, human behaviour isn't adequately explained by modern socio-biology.

I suspect that Darwin himself had doubts about the application of natural selection to the explanation of human high morality. He was deeply troubled by the reception his new theory would receive from the Imperialist, Christian-dominated world view of his native Victorian England. Yet, like any respectable scientist of his day, he reported his findings and gathered his conclusions only in the light of the facts made available to him. Darwin discovered something wonderful and true – he explained that we and all kinds of life on Earth are kin, descended from a single ancestor. He spoke of life as a struggle for existence and of the survival of the fittest. But his *Origins* does not explain the source of aesthetic and moral behaviour in human beings. Indeed, the co-discoverer of evolution, Alfred Russel Wallace (1823-1913), actually held serious doubts about the validity of his theory in explaining the origin of the extraordinary behaviour of human beings. Wallace finally concluded that humankind was an exception to the orderly operation of biological laws. I wonder if this has anything to do with the reason why science has largely ignored the ideas of Wallace while embracing those of Darwin? Why is the theory of natural selection so blatantly synonymous with Darwinism and not 'Wallace-ism'?

Although natural selection may have directed all of life on Earth over billions of years of geological time, human morality is decidedly un-Darwinian. There are some aspects of our behaviour that are actually detrimental to our well-being. Inflicting pain by flailing, fasting or even impaling cannot be written in the genes. For how can they be? All genes know is how to store and transmit molecular information that maintains the life force. And how can the utility of the human spheres of art, literature and music be explained in Darwinian terms? Oscar Wilde was partially correct in claiming that all art is utterly 'useless'. Art and music cannot possibly have a biological role in maintaining the viability of our gene pool. Yet without art and music, the world would be a thoroughly miserable place.

Reflections

The discovery of the DNA genetic code and its manipulation allows us to tinker with the very fabric of our being. We have the ability to alter ourselves, to create, as it were, a new kind of human being. We shall be able to alter any gene at will, allowing us to extend our life span, or perhaps increase our intelligence. Some authorities suggest that all species undergo an inexorable decline. Furthermore, they have pointed out that the selection of particular human beings over others, that is, in *eugenic* practices, allows for the possibility of bringing this decline to a halt. But how can we possibly decide which genes to enrich in our populations? Will we breed brainy intellectuals or super-athletes, or perhaps a mixture of both? Who can possibly decide and by what authority can they act? These are unnerving questions to ask, but the day when we will be confronted by these issues is fast approaching. What we decide will have profound repercussions for the lives and attitudes of our descendants. We shall be faced with a formidable test of our societal adulthood in our emergence from cosmic puberty.

Chapter III

Poised at the Edge of a Biological Cosmos

It is better to debate a question without settling it than to settle a
question without debating it.

Joseph Joubert (1754-1824)

To me, the discovery of an intelligent extraterrestrial life form would
represent the most significant epochal event in our entire history as a
species. In this generation, we have begun to search for such a civilisation
systematically and with astonishing technical sophistication. Why is the
knowledge of the existence of extraterrestrial life so important to us?
What possible changes to our cultures would follow in the confirmation
of such a phenomenon? And where will we find this wonderful and exotic
life? Will it be crawling in the murky shadows of Titan, the planet-sized
world orbiting Saturn, or maybe icy Europa, one of the four so-called
Galilean satellites orbiting the great planet Jupiter? What of the vast potential for habitable worlds around the myriad stars that shine like jewels
in the skies of the blue-green Earth?

Since the dawn of humanity, I can hardly imagine a time when our
ancestors did not ponder on the nature of the Moon and the wandering
planets, and whether there were folk similar to ourselves looking up into
their skies and wondering about the possibility of other sentient creatures
pondering their origins. In historical times, the concept of non-terrestrial
life forms was in itself very dangerous to uphold publicly. Nevertheless,
many philosophers considered the possibility of the plurality of worlds.
Most were persecuted for holding such beliefs. Giordano Bruno (1548-
1600) was burned at the stake for attesting that, amongst other things,
we live in a universe teeming with life. Christiaan Huygens (1629-95),
the great scientific polymath of the seventeenth century, held that the
Earth was but one abode of life and openly speculated on the form and
custom of the inhabitants of the Moon, Mars and even the Sun. It seems
that throughout history, mankind has had a wistful longing to understand
and communicate with beings from other worlds.

The Lure of the Red Planet

Is there life beyond the Earth? I once surveyed a number of my students
with the simple question: 'What is the most likely place to find non-

terrestrial life forms in our solar system?' Almost 90% answered 'Mars'. This is hardly surprising – the planet Mars has fired the imagination of many weighty intellects, past and present. Mars, the fourth planet from the Sun, has been close to the centre of our astronomical attention for many centuries. The ancients watched the Red Planet career across the sky, occasionally looping backwards and then forwards again, as it passed through the constellations of the zodiac. The Romans likened its distinctly reddish hue to the blood-stained colour of the victims of war, and as a consequence named Mars after their god of war. Little else was known about Mars until, as we have seen in Chapter I, Johannes Kepler, using the observational data painstakingly collated by the great astronomer Tycho Brahe, discovered that Mars and the other planets did not move in a circular orbit, but instead swept out elliptical paths around the Sun.

With the advent of the telescope in the early seventeenth century, astronomers were able to peer at the elusive planet Mars for the first time with magnified vision. This magnificent invention also inaugurated the birth of the science of areography, the study of Martian geography. It was Christiaan Huygens who first recorded a surface feature on the planet in 1659. He observed a large, dark, triangular feature, now referred to as Syrtis Major. From careful observations of its apparent movement across the Martian globe, he was able to deduce that the planet had a day slightly longer than ours – about half an hour longer in fact. Mars takes 687 Earth days to make one orbit of the Sun, making a year on Mars roughly equal to 1.88 Earth years. As telescopes improved, so began a sea-change in our perceptions of the Red Planet.

In 1781, William Herschel was able to show that the axis of Mars was inclined about 25 degrees to the plane of the ecliptic – just a little more than that of the Earth. Polar caps had been observed on the planet as early as 1666 by the great French-Italian astronomer, Giovanni Domenico Cassini (1625-1712). It was, however, left to William Herschel to first propose that the Martian polar caps could be made of ice. Here was a planet with an atmosphere and a surface that one could see, having polar ice caps and an orbital inclination that gave it seasons not unlike the Earth. Mars is by far the most Earth-like planet of all the worlds in the solar system.

Swift Little Moons

Mars has two tiny moons, Phobos and Deimos. The story of their discovery is one of the most singular, mystifying episodes in the history of planetary science. It began in 1726 when the Irish satirist Jonathan Swift (1667-1745) described in his book *Gulliver's Travels* the fictional activities of the Laputans, who inhabited an island in the sky. Here is what Gulliver said:

> They have likewise discovered two lesser stars, or satellites, which
> revolve about Mars, whereof the innermost is distant from the centre

of the primary planet exactly three of its diameters, and the outer-most, five; the former revolves in the space of ten hours, and the latter, in twenty-one and an half; so that the squares of their periodical times are very near in the proportion with the cubes of their distance from the centre of Mars, which evidently shows them to be governed by the same laws of gravitation, that influences the other heavenly bodies.

It was not until more than 150 years later, in 1877, that Asaph Hall (1829-1907), an American astronomer based at the US Naval Observatory, discovered two tiny sparks of light orbiting Mars with the newly con-structed 0.66-metre Alvan Clark refracting telescope. These tiny satellites were later named Phobos and Deimos, Greek for 'panic' and 'fear'. Their orbital periods were found to be 7.6 hours and 30.3 hours – within 25% and 40% of the values given by Swift! How did he manage to guess so well?

Swift may have arrived at his conclusions by noting other established facts about Mars. For example, Venus was never observed to possess a satellite, while the Earth had one, and Jupiter was then believed to have just four. Mars, being in between the Earth and Jupiter, would have to have two if a supposed pattern was to be consistent. Swift may also have realised that Martian moons would have to be very small and close to the planet to have avoided detection by Earth-bound telescopes. Just how he came up with the figures for the orbital radii is still something of a mystery and can only be regarded as an extraordinarily good guess.

What would it be like to see Phobos in the Martian skies? Overhead, through the tenuous chill of the Martian air, we would see Phobos rising twice a day in the west and setting in the east. We would see it glimmering through the dust-laced atmosphere, with a brightness several times that of Venus as observed from the Earth. Some day soon, human eyes will behold this celestial phenomenon from the rock-strewn surface of Mars.

Although several spacecraft from both the former Soviet Union and the United States had already been dispatched to study Mars in the late 1960s and early 1970s, none had managed to image the surface features of the small Martian satellites. When the Viking probes arrived in the summer of 1976, all this changed. The first tapes containing the imaging data were apparently disappointing. No clear features of Phobos were discernible, but by repeated image processing, the satellite was shown to be pock-marked by craters. The Viking images also revealed a remark-ably huge crater on the Martian moon, giving Phobos a vague resem-blance to a lumpy sweet potato. It is now believed that these tiny Martian satellites are captured asteroids from the nearby asteroid belt.

Canals and Mars Mania

In the late nineteenth and early twentieth century, an eccentric amateur astronomer founded an observatory at Flagstaff, Arizona, for the sole purpose of studying the planets, above all Mars. This maverick gentleman was Percival Lowell (1855-1916), and his conclusions were to shock the world. Lowell was born into relative affluence, his father being a successful business man. Although he graduated in mathematics from Harvard University, his brief acquaintance with planetary astronomy was more than enough to spark his lifelong love for Mars. He was inflamed with excitement by the findings of the Italian astronomer Giovanni Virginio Schiaparelli (1835-1910), who recorded what he described as 'canali' on the surface of the Martian globe. Schiaparelli was using a fine 0.22-metre refracting telescope which provided excellent images of the planet especially when Mars was closest to the Earth. These fine markings on the planet can only be seen under conditions of excellent observation when the air is exceptionally clear and free from turbulence. 'Canali' literally translated from Italian means 'channels'. Lowell, however, interpreted the features as canals, immediately invoking an intelligent or artificial origin for these phenomena. Although Schiaparelli never conjectured as to the nature of these markings, Lowell's announcement led to a frenzy of rejuvenated interest in the Red Planet.

Mars mania ensued. Lowell supervised the installation of a fine 0.61-metre Clark refractor at Flagstaff, Arizona. This was one of the most powerful telescopes in the world at the time and, being a refracting telescope, was particularly suitable for planetary observation. It seemed the perfect instrument for studying the delicate features of the Martian surface. In the chill of the Arizona mountain air, Lowell painstakingly recorded an intricate network of canals on the Martian globe. They seemed to extend from the polar caps at the north and south to the equatorial and temperate regions of the planet. Lowell conjured up a world of water-craving Martians who were older and wiser than humanity; beings who constructed large and elaborate canals to direct the dwindling water supplies from the polar caps to the barren wastes of the hot, dry deserts at lower latitudes.

Lowell popularised these ideas and continued to speculate wildly in his book *Mars as the Abode of Life*. Contemporary experts on Mars took Lowell's hypothesis with, at best, a pinch of salt. The physical evidence at the time was not consistent with the Martian world conjured up by Lowell. Doubtless, he was a fine and meticulous observer and made many valuable cartographic contributions to the study of Mars. He accurately recorded many phenomena that we know to exist today on Mars. Some astronomers offered their own, more prosaic explanations for Lowell's canals. The legendary planetary observer Edward Emerson Barnard (1857-1923) suggested that had the focal length of Lowell's telescope's objective lens been slightly longer, the canals would have resolved themselves into a series of fine dots. However, other observers using instruments equal to, if not surpassing, the quality of Lowell's telescope, reported no such canals. Others confirmed Lowell's observations but used instruments of inferior quality to those at Flagstaff, and so their results had to be subject to scepticism. The fact that no canals could be seen by other professional planetary astronomers weighed heavy on Lowell's mind, but he remained unwavering in his belief in Mars as a habitat for intelligent beings. And he did not yield to scepticism – Lowell doggedly upheld this incredible story of a water-craving Martian civilisation until his death in 1916.

Science Versus Imagination

As the twentieth century dragged on, the evidence against Lowell's incredible ideas kept steadily rising. Spectroscopy showed that the Martian atmosphere was composed largely of carbon dioxide and smaller amounts of molecular nitrogen and argon. Furthermore, the atmospheric pressure was estimated to be about 30-50 millibars – only 3-5% of our own air pressure at sea level. The polar caps were thought to be chiefly composed of water ice, although dry ice (solid carbon dioxide) was also detected in appreciable amounts. Infrared measurements of the Red Planet revealed that Mars was colder than the Earth with temperatures varying from a relatively balmy 20°C in the equatorial regions to a bitterly cold minus 140°C at the poles. Plainly, Mars was not an inviting place. No evidence for liquid water was seen in any of the spectral data obtained during the 1950s and 1960s. The tenuous atmosphere and the bitterly cold temperatures conspire to make the surface of Mars an astonishingly arid, frigid and hostile place – rather like a bad day in the dry valleys of Antarctica.

During the 1950s and 1960s, interest in Martian life underwent an extraordinary rejuvenation. Many astronomers reported a curious seasonal change in the dark areas on the Martian surface. It appeared that such areas changed colour from a blue-green to a tawny-brown hue as the Martian seasons slowly changed from summer to autumn. A similar observation of the polar caps revealed a regular, seasonal pattern of expansion towards the equator at the approach of the Martian winter, followed by a hasty retreat of the ice caps at the start of the Martian spring. What phenomenon could adequately explain these observed occurrences on the surface of Mars? The consensus among the astronomical community at this time supported the idea that some form of vegetation was present on the planet. Even with the benefit of hindsight, this was an attractive hypothesis and was worthy of some consideration.

The popular astronomy books of the 1950s and 1960s openly portrayed Mars as a planet replete with lichens and other hardy plants. But vegetation should also send a spectral signature across the great void of interplanetary space. If Martian plant life were like the familiar greens that we so fondly cultivate and harvest on Earth, then the truthful eye of the spectroscope should reveal characteristic absorption patterns similar to that of chlorophyll – the green molecule responsible for harvesting the light energy from the Sun. It would make sense for this pigment to absorb wavelengths of light corresponding to the region of the spectrum in which the Sun best emits light – just as happens in the case of terrestrial plants.

Many tried to find the spectral signature of a chlorophyll-like molecule. And although a few scientists believed they could detect tell-tale signs of organic molecules, the vast majority found not a trace of anything remotely organic. Maybe the spectral detection of chlorophyll-like organic molecules was simply beyond the capabilities of ground-based instruments. This seemed very unlikely given that planetary spectroscopy had by this time advanced sufficiently to make these absorption patterns well within the limits of detection. After all, scientists had recorded the thin, carbon dioxide-dominated air enveloping Mars and the distinct signature of limonite in polarised light. Another possibility was that Martian plant life had a very peculiar biochemistry and with our rather blinkered, parochial and incomplete conceptions of biology, we were simply not looking properly.

The meteorological conditions of Earth and Mars were sufficiently different to allow for such a possibility. But scientific opinion gradually swayed away from a biological origin for these Martian features. In the 1970s, Carl Sagan (1934-96) was the first to propose that strange changes in the coloration of the surface markings might be explained by the migration of huge quantities of dust associated with the cycle of the long Martian seasons. Perhaps there was a previously unconsidered change in the exotic Martian soil chemistry? No one could tell for sure. The consensus of opinion was that we had to get a closer look – we had to dispatch spacecraft to Mars!

Viking Invaders

The technology to send spacecraft to the nearby planets developed in the early 1960s, and by 1976 human beings had successfully landed two robotic emissaries to examine a little of the Martian terrain. The Mariner missions to Mars in the 1960s showed definitive evidence that in some ways, the surface of Mars was more like the Moon than an abode of life. Thrilling data dispatched by the Mariner 4 in July 1965 revealed that the planet was puckered by craters, huge extinct volcanoes and enormous canyons and valleys. Many scientists, including the renowned planetary astronomers, Eugene Antoniadi, E. E. Barnard and G. P. Kuiper had predicted many years before that craters would be found on Mars, and Barnard may even have glimpsed the larger craters on the Martian globe under conditions of excellent observation.

The vegetation hypothesis was finally put to rest in 1971. While

Mariner 9 was travelling to Mars, a huge global dust storm enveloped the planet and obscured all surface detail. It was immediately clear that the changes in colour were not due to life but resulted from the generation and propagation of planet-wide dust storms causing rapid changes in the surface features of the planet, as revealed by telescopic observation.

Winds are caused, in part, by pressure gradients – slightly different air pressures arising in different air masses. Wind speed and direction are influenced by the rate of movement of molecules of air going from a region of higher pressure to a region of lower pressure, as well as the rotation of the planet. We on Earth are fortunate in that our atmosphere does not suffer from large variations in air pressure. Fluctuations are frequently no more than several millibars, to an average of about 1,015 millibars. As a result, the Earth's atmospheric pressure changes by only a few per cent. For example, Britain's highest recorded air pressure is 1,055 millibars – just five per cent above normal.

Mars, on the other hand, suffers from dramatic variations in local air pressure, due to its abundance of carbon dioxide. Carbon dioxide is an unusual substance in that as temperatures rise it goes directly from a solid to a gas without going through the liquid phase. This process is called *sublimation*. On Mars, the sudden sublimation of dry ice into the atmosphere at the end of the Martian winter causes dramatic changes in local air pressure: a stunning 33% increase in atmospheric pressure by the end of the Martian summer. In other words, Mars literally has one third less air in winter. But what a beautiful sight it would be to witness a Martian dawn, with ghostly mists of freshly evaporated dry ice gliding over a distant hillside, or across the gently undulating Martian lowlands. These dramatic air pressure gradients dissipate over the entire planet and last for months. On its surface, violent storms rage, reducing visibility to almost zero.

By the early 1970s, it was clear that Mars could not be the abode of intelligent beings with an affinity for civil engineering, as we had hoped. Our perceptions of the Red Planet were to be changed even more dramatically with the data dispatched to Earth from the Viking missions in the summer of 1976. While two orbiters mapped and photographed the planet down to a resolution of 1 kilometre, two semi-intelligent robot landers were deployed to analyse the terrain, atmospheric conditions and soil chemistry of the Martian surface. On 24 July 1976, Viking 1 soft-landed on the gentle, rock-strewn plains of Chryse. Eight thousand kilometres away, on 10 August, Viking 2 touched down safely at Utopia. The photographs sent back to Earth were stunning. They revealed a rusty red desert world, with ordinary-sized rocks and sand dunes studding the landscape all the way to the horizon. Over a period of months, the Viking landers showed us that the Martian temperatures could vary from 20°C on a warm summer's day near the equator to an inhospitable minus 140°C on winter nights. The air was shown to be over three times thinner than expected, composed chiefly of carbon dioxide (95%), with smaller amounts of argon and nitrogen and only a trace of oxygen – in stark contrast to our atmosphere which is rich in nitrogen and oxygen. The first pictures of the Martian sky perplexed the spacecraft controllers back on Earth. The colour image showed that it was decidedly pink and not blue as had been generally expected. It became evident that large amounts of dust present in the Martian atmosphere scatter the Sun's rays far more effectively than the tenuous air molecules, giving the Martian sky a pinkish tinge.

Signs of Life?

The most challenging duty assigned to the semi-intelligent Viking robot landers was to analyse the chemical composition of the Martian soil and determine whether there was microbial life present. Slowly, the robots lowered their purpose-built robotic arms into the red, dusty sands of Mars and retrieved samples that were systematically analysed inside the craft. Three biology experiments were performed in all. The first examined the organic composition of the Martian soil. To the surprise of many scientists, not a trace of organic matter was evident. It seemed that, at least at these locations, the Martian surface material contained no recognisable biomolecules. Next, the analyser mixed radioactive carbon dioxide with the soil in the hope that any Martian plants might 'fix' a little of this gas, in the same way that terrestrial plants do when they convert carbon dioxide and water to sugars in the dark reactions of photosynthesis, releasing oxygen as a waste product. As a control on the experiment, the procedure was repeated on a sample of Martian soil that had been subjected to intense heating. The results were tantalising: something in the Martian sample was capable of chemically combining the gas with the soil, while the control sample yielded wholly negative results.

Was this the eagerly sought-after evidence for microbial life on Mars? Nobody could be sure because chemical analysis of the same soil showed no traces of organic molecules – the stuff of earthlings. In the final experiment, a nutrient broth was mixed with the soil in the hope that any Martian bugs would metabolise it and release waste gases as a token of their humble appreciation. What happened? To the amazement of the controllers back at the Jet Propulsion Laboratory (JPL) in Pasadena, gas was rapidly evolved from the sample as if something in the Martian soil was rapidly breaking it down and subsequently releasing waste gases. But careful study showed that the gas was released very quickly and not gradually as one might expect living, growing organisms to do on Earth. Moreover, the release of gas suddenly ceased after a few minutes, suggesting that it was of non-biological origin. What was going on here? In short, we are unsure. The same results were obtained by both landers separated by 8,000 kilometres. Whatever the explanation for these bizarre results, it was likely to be a planet-wide phenomenon.

The Viking lander confirmed that the soil is composed of an oxide of iron as well as a small amount of highly reactive peroxides. The latter could release oxygen when mixed with water. This could explain the results of the carbon dioxide fixation experiment carried out by the Viking landers. Peroxides are also well known for their ability to be destroyed by heat. Could the heat lability of one or more components of the Martian soil explain the negative results obtained from the carbon dioxide-fixing control experiments? Again, we are unsure. The scientific consensus is that the Viking biology results are not consistent with terrestrial life and that they are best explained in terms of an unusual soil chemistry.

Pathfinding

The highly successful Pathfinder mission to Mars in the summer of 1996 revealed a gold-mine of new information on the geology and atmosphere of the Red Planet. On 4 July 1996, the Pathfinder lander probe, encased in a protective aeroshell, penetrated the Martian atmosphere. As the probe approached the surface, giant air bags inflated all around it, cushioning it from the impact when it finally hit the ground at about 50 kilometres per hour. Electronic monitors in the probe recorded that the craft bounced at least fifteen times before coming to a standstill at a site called Ares Vallis near the Martian equator. Once firmly on the surface, the air bags deflated and Pathfinder could begin its surface operations.

The Pathfinder mission was designed to study the geology and atmosphere of Mars but not to look for life. Using an instrument called a multispectral camera mounted on board the stationary lander, Pathfinder transmitted a wealth of new information on the make-up of Martian rocks and soil. The other part of the mission was carried out by Sojourner, a small, mobile rover carrying a chemical analyser to sniff out the composition of Martian rocks. Pathfinder revealed the Martian soil to be deficient in silicon but richer than Earth's soil in metals like magnesium and iron. What's more, not all Martian rocks appear to be volcanic. Many samples appear to have been formed in sedimentary processes requiring liquid water. Most importantly, a number of smoothed pebbles and mineral conglomerations were found at the landing site, suggesting that they were carved by flowing water in the distant past.

As for the weather, the Pathfinder data showed somewhat similar results to Viking. In the early afternoon, dust devils were photographed sweeping across the lander. But in general the prevailing winds were light – less than about 40 kilometres per hour. Mars, it seems, has not changed much in over 20 years.

The Cloudy World of Ancient Mars

The Viking orbiter images, as well as those from the more recent Mars Global Surveyor, have provided compelling evidence for ancient river valleys, shorelines and countless dried-up tributaries and channels that were apparently carved out by running water. Planetary scientists think that Mars was a much warmer and wetter place about 4.0 billion years ago. This, incidentally, is about the time when life on Earth supposedly arose from non-living, complex organic matter. In this ancient epoch, cometary fragments and other 'left-overs' from the primordial solar system were colliding with the planets and their satellites. Comets contain much water ice and carry a lavish inventory of organic molecules – two vital ingredients for the origin of life. There is no reason to suppose that all the organic-rich material was deposited on the Earth because there was no obvious selection mechanism. Certainly, cometary fragments also delivered the same water and organics to the young Martian surface.

In order to maintain liquid water on a planet's surface a number of physical conditions must prevail. The two most important factors are atmospheric pressure and temperature. Water boils at a lower temperature when the pressure is lower, as any one who has struggled to make a decent cup of tea at high altitude will know. The atmospheric pressure is lower at higher altitude and as a result, water molecules find it easier to escape from the liquid surface. But on Mars the air pressure is so low that liquid water would evaporate furiously, condensing as pellets of ice on the surface. Clearly, the early Martian environment must have had a denser atmosphere – an atmosphere massive enough to give water a chance to flow.

What about temperature? Today the average global temperature on Mars is about minus 40°C, compared with the relatively balmy 15°C global average for the Earth. If the surface of early Mars was not significantly warmer than it is today, water could never have flowed over its rock-strewn surface. But how can an arid world like Mars ever have sustained the right temperatures and pressures to allow liquid water to be maintained on its surface?

The first attempt to resolve the Martian water conundrum was initiated in the early 1980s, when scientists developed computer models of the early Martian atmosphere. In particular, the work of James Kasting, a planetary scientist at Pennsylvania State University, together with more recent work by the astronomers François Forget, at the Marie Curie Institute in Paris, and Raymond Pierrehumbert, of the University of Chicago, has shown that the early Martian environment may have had a dense carbon dioxide atmosphere – up to five times more dense than the Earth's atmospheric pressure at sea level. But a massive atmosphere alone was not the redeemer of the Martian water. Clouds made up of carbon dioxide and water ices may also have been necessary to trap more of the Sun's heat, bringing temperatures up to the melting point of water ice.

How long did the waters flow on our sister world? A difficult question. Conservative estimates suggest that liquid water may have existed on Mars for several hundred million years. If so, then there is a distinct possibility that at least the earliest steps towards life on Earth were taken in this timescale. And given the rapidity with which life on Earth established itself – in less than 100 million years – the odds are again tipped in favour of ancient Martian beasties.

But Mars never had the same advantages in life as big sister Earth. For one thing, it is considerably smaller than our world, and so perhaps its pull of gravity was not enough to keep the gas molecules in its atmosphere from escaping into space. Ultraviolet light is an excellent energy source for dissociating gas molecules. A single photon of high-energy UV light is more than capable of splitting a water molecule into a hydroxyl radical and a hydrogen atom. This type of process is known as *photodissociation*. Since the products of photodissociation are always smaller and simpler than the parent molecule, it follows that the former can escape more easily from the gravitational field of the planet. The Earth is graced by the presence of a minor atmospheric constituent, ozone, which provides protection from the Sun's more lethal ultraviolet radiation. Mars, on the other hand, may never have had an adequate ozone shield, and so it is believed that a runaway destruction of water molecules contributed to the dehydration of Mars over the aeons.

But Earth did not have this problem. Because it is nearly ten times as massive as Mars, our world could hold on to its primordial atmosphere to sustain the complex biochemistry necessary for the emergence of life and eventually intelligence. As a result of the deficiencies Mars faced, its surface atmospheric pressure slowly dwindled to the tenuous levels of the present day. With a diminished atmospheric cover, as the atmospheric gases escaped from the planet any lakes, rivers or seas that may have been present on the Martian surface must have dried up. Mars also lost its ability to keep temperatures above the freezing point of water. This is why many planetary scientists now consider Mars to be locked in a permanent ice age. The interesting question is this: did Mars ever sustain geological conditions suitable for life to emerge in the same way that it is believed to have arisen on Earth? If so, did this early life endure over the aeons of Martian geological time or was it callously snuffed out by the changes that were occurring in the Martian climate?

One (albeit unlikely) possibility is that life may actually have arisen on Mars and then found more favourable conditions on the larger, warmer and denser world that is planet Earth. This cannot be discounted because we have pieces of Mars right here on Earth! For example, there are the SNC meteorites (after Shergotty, Nakhla and Chassigny – names of the places the first specimens were recovered) that have been found in Antarctica over the last few decades. We are certain that they originated from the Martian surface because chemical analysis has shown that gases trapped inside these curious meteoric fragments have an identical composition to the accurately known chemistry of the Martian atmosphere, revealed to us by the Viking lander and Pathfinder missions. The most likely explanation for their presence on Earth is that a large object collided with Mars, ejecting vast amounts of material high into the Martian atmosphere. Most of this material would have settled back onto the planet's surface, but some smaller fragments could have attained a speed equal to or greater than the Martian escape velocity, and as a consequence were lost in interplanetary space, later to be captured by the Earth's pull of gravity. It is also possible (but far less likely) that the opposite occurred: an impact with the Earth may have released biological material into space, which was later captured by Mars. But if this is so, we would expect extant life forms on both Mars and the Earth to have very similar biochemistry, consonant with a singular origin of life.

It is just within the bounds of possibility that life may have been nurtured on many worlds in our solar system. The persistent, cataclysmic bombardment of the entire solar system during the first billion years of its history may have led to impacts that were sufficiently energetic to expel small quantities of biological material into interplanetary space. Many of the terrestrial worlds such as Mars, Venus and Mercury, as well as the more distant large satellites of the Jovian and Saturnian systems, may have been seeded with life at one or more times in the past. This incredible suggestion has been advanced by the geologist Jay Melosh, who has contemplated the implications of such a cross-fertilisation of worlds with microbial life. Melosh suggests that the terrestrial planets should not be considered as being biologically isolated from one another. On the contrary, he argues that these worlds act as a single, complex, ecological niche made up of insular planetary bodies. Even in our neck of the cosmic woods, Melosh argues, life might well be a multi-planet phenomenon.

Interest in Mars, at any rate, has also received a welcome boost from the discovery of the complex dependence of microbes on inorganic material very deep in the Earth's crust. These so-called *lithotrophic* organisms live on the inorganic materials that make up rock. These hardy creatures never see the light of day and appear to grow and reproduce quite happily hundreds of metres below the Earth's surface. The traditional view that all life on Earth is dependent in some way on the Sun is confounded by these rock-loving creatures. There is currently a resurgence of research interest in the study of organisms that thrive in hot springs on the Earth's surface or deep in her oceans at the mouth of volcanic vents that spew out tonnes of energy-rich minerals and gases from the Earth's interior.

No organisms that we are familiar with could withstand the colossal pressures and distinctly unfriendly temperatures of the deep oceans. Yet the microbial organisms that dwell here can easily convert the energy-rich minerals from volcanic emissions into readily utilisable forms of metabolic energy. Although it does not have the elaborate system of plate tectonics that the Earth has, the interior of the Red Planet should also release periodic outflows of these energy-rich substances. If life has evolved to at least the required sophistication to extract the energy from these planetary outgassings, then it is conceivable that these hypothetical organisms will be ubiquitous on Mars and are awaiting our discovery. But not all scientists believe this will be the case. Indeed, many planetary scientists believe Mars always was and always will be a lifeless world. Arguably the most scalding attack against the contention that Mars has life comes from the ideas of the English physicist James Lovelock (b. 1919), co-founder of the *Gaia hypothesis*, named after the Greek goddess of the Earth. Gaia theory sees the Earth as a vast interconnection of organisms which regulates the global properties of the Earth's biosphere. The Earth's atmosphere is rich in nitrogen and oxygen; in contrast, Venus and Mars have very similar atmospheres mostly made up of carbon dioxide.

As discussed earlier, oxygen is a very reactive gas and, if left to its own devices, would quickly react with minerals in the Earth to form oxides. Thus, if oxygen were not continually replenished by the metabolic activities of green plants, its level in the atmosphere would rapidly attenuate. Because of this need for the continual replenishment of oxygen by life, the atmosphere is said to be out of equilibrium, in a kind of constant state of mixing. What's more, the Earth's atmosphere also contains traces of methane and other hydrogen-rich gases, which readily react with free oxygen. The co-existence of these gases with oxygen is most easily explicable if life is present on Earth, which indeed it is. In contrast, the compositions of the Martian and Venusian atmospheres are exactly as one would expect if life were not present. They have atmospheres that are at equilibrium or very close to equilibrium.

Lovelock claims that one does not need to travel to a distant planet to look for life. We need only study the planet's atmosphere to see if any of its constituents are out of equilibrium to deduce a biological presence. But Lovelock's interesting ideas might not be as all-encompassing as he would have us believe. Although it may well ring true for biospheres established on the surface of a world, it would be much more difficult to detect if life forms were present below the surface. Instead of reacting with the constituents of the Martian atmosphere, the metabolic products of life might only modify minerals beneath the surface, leaving the atmosphere largely unchanged.

Meteorites!

Back to the SNC meteorites. On 6 August 1996, a team of NASA scientists working on an alleged Martian meteorite found in Antarctica, announced that they had unearthed physical evidence for past life on Mars. I was electrified when I heard the first news of this revolutionary discovery on a late evening program on television. At the centre of the controversial claim is the prosaically named ALH 84001, a 1.9kg meteorite collected from the Alan Hills region of Antarctica in 1984.

According to an international team of scientists, ALH 84001 was blasted off the Red Planet about 16 million years ago, to be later captured by the Earth only 13,000 years ago. How can we tell how long it spent in open space? During its long, lonely sojourns through interplanetary space the meteoroid is continually bombarded by high-energy particles streaming out from the Sun and beyond, called *cosmic rays*. The rays induce distinct chemical changes in the outer surface of the rock, in a process known to astronomers as *spallation*. By assuming that the rate of cosmic-ray bombardment has remained roughly constant over the last few hundred million years, it is possible to calculate the time a meteoroid has spent travelling through interplanetary space before landing on the Earth.

We are certain the meteorite is of Martian origin because of the chemical composition of tiny bubbles of air trapped deep inside it. When the chemical composition of this trapped gas was compared with the results of the Viking atmospheric samplers, they were shown to be nearly identical. So by studying these trapped bubbles of gas we are releasing ancient and tiny pieces of the Martian atmosphere. This distinct Martian signature has also been confirmed by NASA's spectacularly successful Pathfinder mission.

We cannot be as certain about whether or not the meteorite contains evidence of extraterrestrial life, but there are several indications. Deep inside the meteorite, for example, a group of scientists including geologists, microbiologists and chemists found high concentrations of complex organic matter in the form of compounds called *polycyclic aromatic hydrocarbons (PAHs)*. This organic material could have been derived from a contamination of the meteorite with terrestrial organic material. So how are we sure these organic materials didn't derive from terrestrial sources? One clue comes from the way the organic material is arranged inside the meteorite. In particular, if the organic material comes from terrestrial contamination, we would expect its concentration to decrease as we take samples from deeper within the meteorite. This is in stark contrast to the results found, which indicate that the organic content becomes more concentrated as we probe deeper into the meteorite, strongly suggesting that it is endogenous to the rock.

This sounds promising, but unfortunately it is difficult, if not imposs-ible, to disentangle the origin of the PAHs that have been associated with ALH 84001. These molecules, although complex, can be generated by non-biological as well as living processes. PAHs have even been de-tected in interstellar space. With our present knowledge of biochemistry, it has become possible to elucidate the metabolic origin of many stable organic molecules associated with fossil material. By studying how large, complex biomolecules are pieced together, we can reconstruct the likely events that shaped their synthesis. This new field of *molecular palae-ontology* promises to provide us with novel insights into the primitive chemistries that characterised the first cells on Earth. Unfortunately, PAHs are notoriously difficult to analyse in this way because they are normally formed under conditions of high pressure and temperature (they are com-mon constituents of coal and oil). In fact, you may have unknowingly seen PAHs as constituents of the black, tarry substance on the surface of barbecued meats. Plainly, until we find other species of organic mol-ecules, we will not know for sure whether these PAHs truly were de-rived from Martian microbes.

There are other clues, however. Chemical analysis of ALH 84001 has also identified deposits of carbonates, for one thing. The carbonates of the Earth, which impose themselves so majestically on the eye in the form of the White Cliffs of Dover and also exist as vast deep-sea deposits, are laid down by tiny, microscopic organisms called *formanifera*. These simple cells used calcium carbonate as a major component of their eleg-ant shells. Thus, carbonate depositions could be an indicator of life.

But there are also problems with the Martian carbonates. For one thing, some geologists maintain that the observed carbonate depositions could be explained only by a physical process operating at temperatures in excess of 500°C – too hot for them to be biological artefacts. More recent work by researchers at the Open University in England has, how-ever, challenged this view, by pointing out that the carbonate deposits are, in fact, entirely consistent with a formation temperature below 100°C. Moreover, they have also found evidence showing that the interior of the meteorite contains water-rich minerals, suggesting that they were at one time in contact with running water.

Many terrestrial bacteria, for reasons that are as yet unknown, accu-mulate grains of iron oxide inside their cells as so-called *inclusion bodies*. Another clue to the presence of life, investigators discovered similar depositions co-localised with the carbonate and PAH depositions in the meteorite. Although these depositions can be explained by purely non-biological processes, the evidence on the whole strongly suggests that they were derived from biological material. For example, these iron oxide depositions were found near structures resembling cells. However, microscopic analysis shows that the sizes of these structures ranged from 0.02 to 0.2 micrometres (a micrometre is one millionth of a metre), while terrestrial cells typically fall into the range of 0.5 to 20 micrometres. Some critics feel that these putative 'nanofossils' are thus too small to hold enough genetic material and metabolic machinery to be considered as cellular life forms. But how can we expect Martian life forms to conform to our rather parochial expectations of biology? After all, there are cells in existence on Earth which are only about 0.2 micrometres in size – the upper limit for the dimensions of the putative Martian fossils. Is it not conceivable that smaller cells will be found on Earth, perhaps lying mid-way in size between a virus and a regular bacterium?

Another meteorite of undisputed Martian origin has also been shown to be extraordinarily rich in organic compounds. This piece of Mars, named EETA 79001, has been known since the early 1990s to harbour organic molecules, but this was largely ignored because of suspicions that they were derived from terrestrial contamination. The British scientists conducting the investigation into EETA 79001 have estimated that the total organic content of the meteorite is of the order of 0.2%. Preliminary analysis has also shown that a major constituent of this organic material is amino acid-based. As we have seen, amino acids are the building blocks of proteins, the slaves of terrestrial cells. Unlike ALH 84001 which is 4.5 billion years old, this second meteorite appears to have formed only 180 million years ago, and was blasted off the Red Planet about half a million years ago. Because the conditions on Mars (or on Earth for that matter) have hardly changed at all in the last 180 million years, and if these organic molecules were indeed generated by biological processes, then life may still exist on our sister world.

Since the publication of the Martian nanofossil paper by the NASA scientists in 1996, there has been a flood of new research interest in these meteorites. And while there is as yet no consensus of opinion on the matter, it seems clear that the only way to settle the question is through a thorough search conducted by humans on Mars. With the race to place a human on the Red Planet steadily gaining momentum, we will soon know the answer.

Life as a Cosmic Phenomenon

If the cosmos were a mirror, what would the Earth reflect? Our Sun is an ordinary star, past the uncertain times of its youth. It is one of billions of other Sun-like stars scattered across the Milky Way Galaxy. Most stars have stable planetary systems, ripe for our exploration. And although we have not yet found them, we passionately believe that the Galaxy should be pregnant with myriad exotic, swimming, walking, flying, talking and thinking creatures. Everything we have come to observe about this great and imposing universe imbues us with a sense of our own mediocrity. If the universe is to be self-consistent, it must have spawned life in many locales and in many epochs – past and present.

The English physicist Paul Davies (b. 1946) has advanced three philosophical principles supporting the idea that extraterrestrial life forms exist. These are:

1. *The uniformity of nature:* Simply put, this is the assumption that the laws of nature are the same throughout the universe. As yet, there is no evidence whatsoever to suggest that this is not the case.

2. *The principle of plenitude:* Whatever is possible in nature tends to happen, given enough time. The fact that life arose on Earth therefore strongly suggests that living systems should be found throughout the cosmos.

3. *The Copernican principle:* In the sixteenth century, the Polish astronomer Nicholas Copernicus advanced the idea that the Earth is not at the centre of the universe, but instead orbits the Sun together with the other planets in our solar system. After the verification of this heliocentric cosmology, humanity was dethroned from the centre of Creation. As an extension of the *Copernican principle*, the Sun is now known to orbit the Galaxy about two-thirds of the way out from the galactic centre. Our star is typical in almost all respects. It follows, extending the principle further, that life on Earth may be typical of a plethora of life forms inhabiting the cosmos.

But might there be an as yet undiscovered principle that precludes the emergence of life on a universal scale? And do these principles apply to intelligent life also? Many biologists believe that the emergence of even primitive life on another world is a formidable achievement. To expect to discover extraterrestrial civilisations, with technology vastly superior to our own, many biologists argue, is to expect nothing short of a miracle.

And just as there are principles supporting the existence of extra-terrestrials, so too are there counter-theories. A school of thought has arisen in recent years based on the *Goldilocks principle*, as advanced by George Weterhill at the Carnegie Institute of Washington, D.C. The central tenet of this principle holds that a long line of very improbable conditions had to be 'just right' in order for life and intelligence to emerge and develop. Take the Moon for example. This giant ball of rock orbiting the Earth has proved to be critical in stabilising the Earth's axial inclination, allowing regular seasons to assert themselves. Moreover, if the Moon were significantly farther away than it is now, the Earth's tides would be considerably weaker. Tidal movements have helped to orchestrate the emergence of land animals through the selection of organisms that could withstand longer and longer periods above water. Without tides, the impetus for early animals to conquer dry land would be diminished.

Or consider the location and prowess of the mighty planet Jupiter.

Orbiting about five times further away from the Sun than the Earth, and possessing over 300 times the mass of our world, Jupiter has been responsible for protecting our planet from cometary impacts for billions of years. Because it is so massive, it can deflect comets into erratic orbits before fully expelling them from the solar system. Thanks to this giant sister world, many comets that would otherwise have collided with the Earth were warded off.

These are but two examples of the Goldilocks principle. They seem to be compelling arguments for the unlikelihood of the emergence of intelligent life on other planets, and indeed they are, for the specific case of the Earth! The truth is that we have as yet only a parochial view of the minimum requirements necessary for the emergence of living things. Other planetary systems may not have a natural satellite where the Earth's Moon is, nor may they hold a Jupiter-like world in precisely the same location as it is in our solar system, yet there is a whole host of other physical conditions that may do the job just as well.

Perhaps the most admirable attribute of biological systems is their hardiness. Once life emerges on a world, it adapts rapidly to its changing environment. Given time, it will exploit every available niche presented to it. On some worlds at least, it seems likely that the path followed by life would have made way for the emergence of intelligence, not least because it confers a substantial survival advantage to the organism. The Goldilocks principle may well ring true, but the emergence of life on Earth may be just one expression of its validity, one staging of an ongoing drama.

For me, the whole question of whether life is common in the cosmos hinges on how easy it is for life to get kick-started. Nature rarely develops forms that are singular to the cosmos, with its 100 billion galaxies and 10 billion trillion suns. Why shouldn't this be the case for life forms also? As we have seen in Chapter I, there seems to be a real problem accommodating Darwinian natural selection at the centre of a new, cosmological theory of biology. There are some very puzzling things about the origin of life on Earth – its enigmatic birthplace, the extraordinarily rapid phase of transitions leading to the 'complexification' of living things, and so on – which show that life may have arisen much more rapidly than Darwin's theory can account for. The accumulated wisdom of our observations of the universe paints an entirely different picture – a perception of life that arises spontaneously whenever certain conditions come together.

The Habitable Zone – Debunking a Scientific Myth

In the early 1960s there emerged the concept of a habitable zone – a shell of space around a star within which liquid water and hence life may exist. In 1966, in the classic book *Intelligent Life in the Universe*, the astronomers Carl Sagan and I.S. Shklovskii wrote, "We can imagine each star surrounded by a spherical shell, throughout which planetary temperatures are equable, and the origin and development of life are possible. We can call this the zone of habitability or the ecosphere."

This thought-provoking concept provided the bedrock upon which astronomers have based their ideas about the locales for the emergence of extraterrestrial life. But as powerful as this concept may be, it is very restrictive and possibly even misleading! New research in disciplines as distinct as biological oceanography and astronomy are beginning to paint a new picture – a paradigm shift in our perceptions of life in the cosmos.

Apart from carbon atoms, water plays the pivotal role in the lives of terrestrial organisms. In terms of its physical properties, water is a very special molecule. Composed of two atoms of hydrogen and one of oxygen, water is by far the most abundant molecule in any living cell. Two hydrogen atoms are bonded to an oxygen atom, causing the hydrogen and oxygen atoms to share electrons between one another. However, oxygen atoms have a greater affinity for electrons than hydrogen atoms, and as a result the water molecule takes on a *polar* character – the oxygen atom having a slightly negative charge and the hydrogen atoms a slightly positive charge. The strong polar character of water allows it to engage in unusual bonds with other water molecules – scientists call them hydrogen bonds. Though individually very weak, the myriad hydrogen bonds in water are responsible for some of its magical properties such as its surface tension and unusual stability over large ranges of temperature.

Another unusual property of water molecules is their propensity to expand on freezing. Everybody knows that ice floats on water. This is because it is less dense in the solid state than in the liquid state. Most molecules of similar size and mass to the water molecule increase in density after undergoing the liquid to solid transition. The fact that ice is less dense than water explains why water freezes from the surface down. Think of the consequences if it were the other way round. Ponds and lakes would slowly accumulate ice. As time progressed all the ice would sink to the bottom. Rivers would grind to a halt. Once the water was permanently frozen, all but the hardiest microscopic aquatic life forms would perish. Aquatic life forms would not have survived for long had the properties of the water molecule been different. Moreover, most other solvents that are both abundant and stable tend to be liquids only at very low temperatures. Liquid water, on the other hand, is stable over a very wide range of temperatures and pressures. To sum up, although water is extraordinarily common in the universe, its properties are extremely rare.

Far out in the Saturnian system, there is a moon called Titan. Almost as large as the planet Mars, Titan holds a dense atmosphere composed of nitrogen and the simple hydrocarbons methane and ethane (with a pressure 50% higher than the Earth's air at sea level!) and may have lakes of liquid ethane. The average surface temperature of the bizarre Titanian landscape is a very chilly minus 178°C. At this temperature, just 95 degrees above absolute zero, molecules must move very slowly, perhaps too sluggishly to form complex life forms. Moreover, if there are lakes of liquid hydrocarbons on the surface of Titan, they will only dissolve fat soluble substances. In contrast, water is a remarkably efficient and broad-purpose solvent, capable of dissolving a wide variety of salts and nutrients – necessary ingredients for biological systems.

Our ongoing exploration of the planets in our solar system is, however, revealing more promising abodes for past or present life. Data sent back by the Voyager probes, and more recently by the Galileo Jupiter mission, have given us a glimpse of another water world in our solar system – a world without an atmosphere. One of the Jupiter's four largest satellites, Europa, is comparable to our Moon in size, and has a surface of rock and water ice. High-resolution images obtained by the Galileo Jupiter orbiter have revealed that Europa is criss-crossed by a network of huge grooves and indentations, suggestive of a highly plastic crust. Something below the Europan crust is slowly reworking the surface. The likely explanation is that an ocean of liquid water lies beneath Europa's icy crust.

Because Europa's orbit lies between Jupiter and the other Galilean satellites, it is engaged in a gravitational tug-of-war which is sufficiently strong to flex and heat Europa's interior. These tidal forces cause vast amounts of rock and ice to grate together, generating huge quantities of heat in the process. The energy liberated by this tidal friction may be sufficient to maintain a subsurface ocean of liquid water. With enough organics and the right temperatures, the conditions might be ideal for the development of primitive life forms. Such an oceanic world would be a very dark place. Creatures would have to evolve in very different ways to terrestrial organisms in order to survive. Until we dispatch machines that can probe these alien waters, the question of life on Europa will have to remain an undemonstrable yet tantalising possibility – a possibility that might be confirmed in the coming decades.

The discoveries of the existence of subsurface oceans on Europa and of the cloudy and wet nature of ancient Mars have caused a sea-change in our perspectives of where life can emerge and evolve. Our speculations about extraterrestrial biology are firmly grounded in a 'terracentrism' – the misconception that only worlds similar to Earth in their size and distance from a Sun-like star can manifest life and mind. But even with Mars and Europa, we have not exhausted all the possibilities.

Consider the possible ranges of terrestrial planet masses. In our solar system, Earth holds the distinction of being the largest and densest terrestrial planet, but there are reasons to think that considerably larger planets are, as we speak, speeding round Sun-like stars in the cosmic neighbourhood. They cannot be much bigger than the Earth. We know that Jupiter and the other gas giants in the solar system were formed by the accretion of about five to eight Earth masses of ice and rock before undergoing a runaway growth, sweeping up vast quantities of hydrogen and helium from the proto-solar system. Clearly, if a terrestrial planet reaches this mass, it will most likely evolve into a Jovian-type world, without a recognisable surface. Yet it is distinctly possible that terrestrial worlds between, say, two and three Earth masses may loom large in alien solar systems. What are their chances of developing life? It may well be that these forgotten, (or at least unconsidered) possibilities will force yet another change in our perceptions of extraterrestrial biology.

Larger terrestrial planets would acquire a greater quantity of gases and accrue larger amounts of water and organic molecules from comets. These worlds would almost certainly experience a lot of volcanic activity in their early days, allowing biologically useful gases to be released from the planet's interior. If a world with, say, two or three Earth masses settles down at a safe distance from its sun – well beyond the distance of Mars from the Sun – the conditions may be ripe for the origin and development of life. It might be a world where liquid water oceans cover the surface beneath a great sea of air, exerting an atmospheric pressure several times that on the Earth at sea level. The dense atmosphere would generate a greater greenhouse effect, making it easier to maintain liquid water oceans even at large distances from a sun. For now, these large terrestrial planets remain only a theoretical possibility. But optical technology of the early twenty-first century may well uncover these water worlds following their eternal paths around their parent stars.

We have seen how the concept of a habitable zone is quite misleading when we consider the cases of Europa and ancient Mars. The energy needed to create and maintain a biosphere may not only be furnished by radiant energy from a star, it can also arise from the conversion of gravitational potential energy into thermal energy (tidal heating), or through the establishment of greater greenhouse warming at large distances from a Sun-like star. If Europa turns out to be an abode of life, how many other worlds scattered across the cosmos will also support a bounty of living things? Indeed, how can we say that all life depends on the warmth of a star? There may be giant planets roaming the cold dark of interstellar space, forever denied the warmth of a sun. And if these colossal worlds were to have their own miniature solar systems, as Jupiter does, might life be possible deep beneath the surface of these moons in an ocean of water maintained by tidal friction? The laws of physics and biology say it is a possibility!

But we humans are not just content with discovering insentient extraterrestrial creatures. We have a great, soaring passion to discover other thinking creatures in the cosmos. We seek creatures who think and wonder like we do. We seek beings who can truly appraise the status of humanity in this vast and imposing universe. We are in search of cosmic companionship. How might one start to begin the search for other sentient beings in the Galaxy? The answer, according to most astronomers, is by radio.

The Search for Extraterrestrial Intelligence

As we have seen, radio waves, like all other forms of electromagnetic (EM) radiation, can travel at 300,000 kilometres per second in a vacuum. At this speed, it takes EM radiation only about 12 hours to traverse the entire solar system. EM radiation from radio signals could travel in only a few years to Proxima Centauri, the nearest star to our own solar system, and would reach a great many nearby star systems in the life span of a typical human being. With radio astronomy, the possibilities are immense. It is feasible to send highly complex information, albeit in a cryptic format, to the far reaches of the Milky Way. What's more, sending information by radio is very cheap.

In order to communicate by radio, one must construct a transmitter as well as a receiver. But which wavelength do you transmit or receive at? This has always been a difficult question, and there has never been a consensus of opinion on the answer. Many astronomers consider the universal radio signature of neutral atomic hydrogen as a likely cosmic 'magic frequency'. This is located at a wavelength of 21cm band and represents the feeble energy difference between parallel alignments of the hydrogen atom's nuclear spin and electron spin, and that of the corresponding anti-parallel conformation. Any competent extraterrestrial astronomer will know about the 21cm hydrogen frequency. As we have seen, it has revealed a veritable treasure chest of information on the distribution and physical characteristics of matter in our universe. Perhaps most importantly, the 21cm hydrogen frequency is located in a relatively noise-free region of the electromagnetic spectrum, making it easier to filter out random signals from artificial signals.

And what of the artificial signal itself? How could we tell if it were a mental product of other intelligent beings? Most astronomers think an artificial signal will be transmitted in a very narrow waveband to maximise the power of the signal. What's more, it should have a pattern that any intelligent being would easily recognise. Since mathematics is a universal language, it follows, so the astronomers think, that the signal will be mathematical in nature. But one can never know for sure. They might even be beaming their musical masterpieces across the void of interstellar space.

Since the distances between the stars are immense, and because the nearest technologically competent civilisation may be hundreds or even thousands of light years away, it would take our transmissions several human lifetimes even to reach them. The likely timescales involved in establishing dialogue are even longer than this. For this reason, it makes more sense for us to eavesdrop rather than transmit. This logic has been adopted by *Search for Extraterrestrial Intelligence (SETI)* researchers.

What are the odds of contacting an alien civilisation? Very small indeed. In the 1960s, Frank Drake (b. 1930) formulated a simple mathematical equation to compute the number of extant technologically advanced civilisations in the Milky Way Galaxy:

$$N = R_* \times f_p \times n_e \times fe \times f_i \times f_c \times L$$

Where

R_* = the mean rate of star formation

f_p = the fraction of stars with planetary systems

n_e = the mean number of planets in each system with environments favourable for the origin of life

f_e = the fraction of such favourable planets on which life does develop

f_i = the fraction of such inhabited planets on which intelligent life with manipulative abilities arises during the lifetime of the local sun

f_c = the fraction of such worlds on which intelligent life forms with advanced technology develop

L = the lifetime of such technological civilisations

This is known as the *Drake equation*. Its present form developed as a result of a conference on intelligent extraterrestrial (ET) life, held in November 1961. We can readily estimate numerical data for the first two or three terms of the equation with at least a modicum of confidence, but as we progress through the later terms, the values can only be guessed at, albeit intelligently. Plainly put, we simply do not have enough information to compute with any accuracy the value of N – the number of technologically advanced extraterrestrial civilisations presently in existence in our Galaxy. Therefore, we cannot begin to compute their average distribution in the Galaxy. Pessimistic guesstimates place N at between 10 and 10^6, that is, between ten and a million extant galactic civilisations capable of interstellar communication. Some ardent pessimists put this figure at N = 1; that is, we are the only technical civilisation present in our Galaxy at this moment in time.

The earliest and arguably most famous search for ET was *Project Ozma*, initiated by Frank Drake in 1960 using the newly constructed 24-metre radio-telescope at the National Radio Astronomy Observatory at Green Bank, West Virginia. He examined the sky for signals using just one radio channel. Today SETI researchers can analyse hundreds of millions of radio channels at once. Since the 1960s over 70 small-scale searches have been attempted. Nearly all of them used the 21cm hydrogen frequency. But not all SETI researchers agree on the 'universal logic' of the hydrogen spin flip. The late SETI researcher Bernard Oliver pointed out a relatively noise-free region of the microwave spectrum, corresponding to the emission lines of hydrogen and hydroxyl ions, the dissociation products of the water molecule. Because of the curious nature of this region of the microwave spectrum and its relationship to the search for extraterrestrial life, it has been optimistically named the 'water hole'. Ultimately, however, we may have to search the entire accessible EM spectrum.

In October 1992, NASA funded a SETI program called the *Microwave Observing Project* (*MOP*), which was initially planned to examine 800 nearby Sun-like stars over a frequency range of 1,000 to 10,000 Hertz. But searching for ET in the modern US political climate was never easy and, unfortunately, the US government reneged on its endorsement of SETI within one year of the commencement of the search. However, radio astronomers were unshaken in their determination to carry out a comprehensive search. Thanks to private donations from wealthy individuals and contributions from members of the Planetary Society, a number of searches have been sustained. During the 1980s and early 1990s, the available technology allowed up to 8.4 million channels to be analysed simultaneously. But the continued reduction in the cost of such high-technology devices has now made it possible to design and build systems that can analyse as many as a billion channels at once.

Smaller teams of scientists at Stanford, Harvard and the University of California at Berkeley have initiated their own searches. These and other searches have been adapted to 'piggyback' on radio telescopes. The rationale here is straightforward and prudent. Instead of applying for precious telescope time, which is already set at a premium for other, more conventional 'hard science', the analysers simply monitor incoming radiation from whatever source is on view.

At the other end of the scale is the ambitious proposed SETI program known as *Project Cyclops*. This gigantic engineering feat would involve setting up an array of 1,000 to 25,000 radio antennae spanning an area 16 kilometres in diameter. Such an array would be capable of picking up signals from anywhere in the Galaxy but because of its colossal cost it seems that the project is very unlikely to be realised in the foreseeable future.

Where should we begin to look for messages from our cosmic cousins, if they exist? Instinctively, we would pay close attention to Sun-like stars that are long-lived and rich in heavy elements like carbon, oxygen and silicon – the stuff of worlds and life. In our Galaxy, astronomers have classified most stars into two groups, *Population I* and *Population II*. Population II stars are very old and long-lived but are poor in heavy elements. In contrast, Population I stars are younger, having condensed from recycled matter from the previous generation of stars, and are rich in heavy atoms. As a result, astronomers think that Population I stars are far more likely to harbour planets suitable for the development of life and intelligence. The Sun was formed about 4.55 billion years ago from the processed matter ejected from posthumous stars. But many stars are significantly older than the Sun and are just as rich in heavy elements. The planets orbiting these stars would enjoy a long and stable life – worlds that may provide comfortable abodes for intelligent beings.

Factors Dictating the Emergence of Extraterrestrial Intelligence
The vast majority of biological organisms on Earth cannot be considered intelligent in any conventional sense. Modern science describes much of animal and plant behaviour in terms of instinctive repertoire, based on commands orchestrated from pre-existing genetic programmes. Intelligence, on the other hand, is the use of cognitive powers to attain a similar result. Birds and bats have the genetic know-how to sail the breezes of heaven, but humans have created aircraft through a rational understanding of their environment, allowing motion through the skies with similar proficiency although with less grace. Instinct and intelligence are two different ways of solving the same biological problem.

Intelligent species are far more numerous on the Earth's land masses than in her oceans. Why might this be the case? Could it in any way provide a clue to the environments in which intelligence flourishes? Let's consider first the categories of animals that can be considered intelligent.

It is arguably true that mammals are smarter than reptiles or amphibians. And within the mammalian lineage, there seems to be a gradual increase in intelligence from moles and shrews, through cats and dogs, and at the top, primates such as apes and humans. However, there are very intelligent mammalian species in the oceans. The cetacean mammals, which include whales, porpoises and dolphins, are highly intelligent, emotional beings, who have forged a unique relationship with the waters of the Earth. But one look at the skeletal anatomy of a whale and it is obvious that it was derived from a terrestrial animal of sorts. It turns out that dolphins and whales were originally land-dwelling mammals who, for unknown reasons, re-embraced the sea about 60 million years ago: they too have a land-based ancestry.

The Lure of the Land
I believe that just as intelligence evolved on the land masses of our world, so too it is likely to preferentially develop on the land masses of other terrestrial worlds. Why on land? What distinguishes the land from other terrain on the Earth's crust such as her oceans and atmosphere? One answer is the rate of physical change of the environment. The land heats up more quickly than the oceans, causing greater and more varied seasonal

changes as well as environment-altering climatic changes over the ages. Land animals had to deal with these changes, which would demand alterations in behaviour patterns culminating in more rapid evolutionary development. Moreover, land animals have had to cope with the changes brought on by the Sun's diurnal journeys. Creatures living deep in the sea, where sunlight is already perpetually dim or even non-existent, would be less perturbed by the coming of the night. In short, the utterly new frontier of the land brought fresh challenges and opportunities for life on Earth.

For over 90% of the history of life on Earth, life was confined to the oceans and the world did not manifest a recognisably intelligent species. It was only after the colonisation of the land masses by animals that brain power and intelligence began to flourish. If intelligence were destined in any way to develop in an ocean environment, it had ample chance to do so. There is something about the sea, its comfort and relative change-lessness perhaps, that prevents a technological species from ever emerging from within it.

Those early amphibian-like creatures that made the pioneering first steps to dry land encountered formidable obstacles in their transition from the oceans. For one thing, they must have felt the tremendous weight of the atmosphere, which posed considerable constraints on locomotion. The changing landscapes, together with the vagaries of the Earth's climate, forced some species to figure out the patterns inherent in nature. This natural process eventually led to the birth of the genus *Homo* about two million years ago.

There are even more compelling reasons for thinking that a technological species must be land-dwelling. In order for a creature to develop technology, there must be a ready source of raw materials, such as rocks, sticks, shells and the like. It is easy to see how a technological species can develop on land. There is a readily available diversity of materials from which all manner of tools and gadgets can be fashioned. But in the seas and oceans, most 'handy' material is located on their sable floor, away from sunlight and the vast majority of ocean-dwelling creatures. Only beings eking out a living in this swarthy abyss of the ocean floor would have ready access to technological materials. But our deep-sea expeditions have not so far located a civilisation beneath the waves. So we can cautiously conclude that other such factors prevent the emergence of intelligence in these environments.

But what provides the impetus for intelligence? We have already seen that intelligence most probably arises on land, where the environment changes most rapidly. This is not just idle hearsay. We really do have unequivocal evidence that changing, colourful environments stimulate brain growth. As early as the late 1970s the American psychologist Mark Rosenzweig and his colleagues at the University of California at Berkeley performed a remarkable series of experiments on brain changes during periods of learning. They set up two different populations of lab. rats – one in a dull, impoverished and unchanging environment, the other in a colourful, eventful, variegated environment. When anatomical and biochemical analysis was done on the animals' brain tissue, Rosenzweig discovered that rats brought up in the stimulating environment displayed markedly faster rates of nerve growth and cortical tissue mass-gain compared to their impoverished counterparts. Not only that, the biochemical results accurately reflected his autopsy studies, showing large increases in the rate of mRNA (*messenger RNA*) production in the brains of rats reared in the stimulating environment; mRNA, in turn, gets translated into functional proteins which maintain the majority of cellular activities. These and other experiments clearly suggest, therefore, that changing environments stimulate brain biochemistry and evolution. This mechanism should operate on any habitable world in the cosmos and in any epoch.

But environments may exist on other worlds that change even more rapidly than those on Earth. The eternal gravitational tug-of-war between the Earth and the Moon has significantly changed the length of the Earth day over a surprisingly short timescale. About 200 million years ago, when dinosaurs ruled the Earth, the day lasted only 20 hours. How might the length of a planet's day influence the emergence of intelligent beings? Let's look at one example. Suppose the Earth spun more rapidly on its axis, making the days and nights shorter. On worlds with shorter days, animals would need to develop more organised and integrated cognitive skills to figure out how to maximise food production during the shorter periods of daylight. Natural selection would tend to favour the survival of organisms that could figure out the ephemeral changes of day into night, and plan their activities for the hours ahead. In contrast, worlds that spin on their axis more slowly than the Earth would have correspondingly longer intervals between daylight and dark, relieving the pressure on creatures to figure things out so desperately.

Perhaps this was an important factor in the emergence of intelligence on our world? The genetic 'seeds' of intelligence may have sprouted in a rudimentary form at the onset of the great Cambrian explosion, some 550 million years ago. At this time, the days were even shorter than 20 hours, and this would have imposed strong selection pressure to grow brains. But the Earth and the Moon – the *Double Planet*, as they are affectionately referred to – may not be typical. There may be a great many habitable worlds without relatively large satellites. These worlds may spin more rapidly than the Earth, and over long periods of time. Maybe short-day worlds have very intelligent creatures living on their outer skins!

Where is Everybody?

If there are many technologically advanced space-faring civilisations in our Galaxy alone, why have they not made contact? This is an important question, as it hinges on the significance of the human species in the cosmos. Perhaps the most interesting possibility is that we actually have received hundreds of intelligent signals from deep space, but we are as yet unsure of their significance. Radio astronomers have detected hundreds of signals that seem to satisfy all the criteria needed to confirm the existence of extraterrestrial intelligence, except for one vital attribute – verifiability. None of the signals so far detected have ever shown up for more than a split second before disappearing. Never have we detected a signal that has both the attributes of intelligence and repeatability. Nonetheless, there is an intriguing feature of the candidate signals that is worth mentioning. All of the candidate signals detected by *Project META* (*Million Channel Extraterrestrial Analyser*) were located within the plane of the Galaxy, suggesting that they could not have been terrestrial in origin.

Perplexed by these tantalising signals that flitter out of existence before they can be confirmed, astronomers are now beginning to wonder if, like starlight, artificially generated signals may undergo a flickering or scintillation. Interstellar space is suffused by very weak magnetic fields which have the power to bend and distort any radio signals. This has the effect of causing the signals to wink on and off on their journey between the stars.

Another possible answer is that we have been visited in the past. Think of a time, perhaps very recently in geological terms – say, one million years ago. An alien scout ship arrives in orbit around our beautiful blue-green world. An atmospheric probe is dispatched to study the constituents of the Earth's air and surface features. The probe returns overwhelming evidence of biological activity.

It would be obvious to the visitors that a riot of life exists both on dry land and in the oceans. Perhaps a probe studying the migrations of the vast herds of strange quadrupedal life forms on the savannahs of the African continent would come across a curious, tool-using bipedal life form, *Homo erectus*. The probe reports that this promising animal displays at least a modicum of self-awareness and is able to exploit its environment more efficiently than other life forms on the planet. Perhaps the visitors would consider this creature promising, but unlikely to achieve a level of sentience necessary for communication with the technologically advanced creatures of the cosmos. The scout ship leaves the solar system, unlikely to return. Plainly, our ancestors may not have been aware of such a visit from an extraterrestrial civilisation. In the very short period since then, *Homo erectus* ancestors evolved rapidly. How I wonder what this civilisation would think of humanity now.

But the cosmos is vast beyond belief. A civilisation that emerged slightly earlier than humanity would have no compelling reason to visit this end of the Galaxy – all sectors of the Galaxy would seem equally inviting. There would be no artificial signals emanating from our solar system and our vast repository of biological organisms would perhaps be considered common place in a well-explored region of the Galaxy. It has been widely suggested that an advanced extraterrestrial society would know about our level of existence but, because of some moral obligation, would not interfere with our progress. But there is one other possibility that we cannot, as yet, dismiss. What if we are the only advanced technological civilisation in the universe, or at least in the Galaxy? Some civilisation must be first to emerge in the history of the Galaxy, and humanity may hold that distinction.

The lack of success of radio astronomy in detecting extraterrestrial intelligence has inspired other astronomers to use alternative methods of detection. How else could an extraterrestrial signify its presence in a vast universe? One possibility is that extraterrestrial life may inadvertently reveal itself through the activities of its technologies. One suggestion explored by SETI enthusiasts is that an advanced extraterrestrial civilisation might make use of matter and anti-matter to generate enormous quantities of energy. For instance, the collision of an electron with its corresponding anti-particle, the positron, results in the total annihilation of both, with the generation of pure energy in the form of high-energy gamma rays. Searching the sky for anomalous gamma-ray emissions, some astronomers believe, could be a way to detect the activities of extraterrestrials. Other scientists are working on the idea that extraterrestrials could use very precise laser beams to signify their presence, or as a tool for interstellar communication. This idea was first advanced by the Noble Laureate in physics, Charles Townes (b. 1915), who invented the laser. Yet another imaginative idea proposed by SETI enthusiasts is that ET could alter the chemical composition of its home star by dumping nuclear waste onto its fiery surface. The dissipation of these exhausted nuclear fuels could result in a 'peculiar' stellar spectrum, showing evidence of advanced extraterrestrial technologies. To date, however, nothing has been found.

In the late 1950s, Freeman Dyson (b. 1923), a prominent American physicist, advanced his own ideas about how an advanced extraterrestrial civilisation could be detected. According to Dyson, an advanced race of aliens would be able to harness almost all the energy from its star by constructing an enormous sphere, centred on the star and soaking up all its radiant energy. Such a *Dyson sphere*, as they have come to be called, would heat up and eventually radiate in the infrared. The radiation emitted from such an enormous sphere could be detected over hundreds or even thousands of light years. So far, no candidate Dyson spheres have shown up.

These are but a few of the imaginative ideas put forward to detect alien intelligence in the cosmos, but maybe we're missing the point. Are we correct in the very philosophy we adopt in our search for these beings? How can we be sure that the motivations of extraterrestrials are similar to ours?

The culture of our ancestors has played a powerful role in shaping the way we view our place in the universe today. Biological evolution made way to cultural revolution in the childhood of our species. We have evolved into a knowledge-generating and knowledge-seeking species. Our indefatigable curiosity and belief in our own destiny has made us the custodians of our world. But the cultures developed by extraterrestrials may not be anything like the ones we are familiar with. Let us take a specific example.

Suppose an intelligent species developed on the land masses of a world permanently shrouded in cloud. These creatures would never see the stars. They could build tools and cities. They would have to figure out how to tell time and recognise the passing of the seasons by some method not involving the position of their sun or the stars in the sky. These creatures would never develop a natural curiosity for the stars, because their culture would never absorb the marvels of the celestial sphere. These creatures would, sooner or later, discover the atom, Newtonian physics (by some other name of course!) and the principles of quantum mechanics. They might eventually figure out that *something* lies beyond the clouds, but without knowledge of a vast and ancient universe – information that could be readily gleaned from just observing the stars – this extraterrestrial society would be severely set back in their motivations for exploring the cosmos.

Think of how different we might have been had we lived on a world where the daytime glory of the Sun and the lesser splendour of the stars in the night sky were never visible. These great astronomical objects have been at the centre of our cultural and spiritual inclinations for hundreds of millennia. Our ancestors patiently watched and recorded the motions of the Sun, Moon, planets and distant stars, with the passing of the seasons. Much of the religion and myth of the human species is centred on celestial objects. The Sun and the stars have also profoundly influenced our mathematical discoveries. The rising and setting of the Sun and other stars give the impression of a great circle in the sky – a celestial sphere. The near-circular faces of the Sun and full Moon may have also inspired some mute and forgotten mathematicians of prehistory to contemplate the geometry of the sphere and circle.

What's more, many of our concepts of the nature of light and other forms of electromagnetic radiation have been inspired by studying the stars. Even some of the elements were discovered not on Earth, but in the Sun. As we saw in Chapter I, helium was discovered from an analysis of the Sun's spectrum. Indeed, other species on the planet Earth have developed a deep affiliation with the stars. Some birds, such as migratory geese, navigate by using the stars. They can identify the positions of many of the bright stars in the sky and can apparently compensate mentally for their changing positions through the night.

It is clear that in these and many other ways, the knowledge of the stars has powerfully influenced the evolution of many of the species on our planet. Nowhere is the influence of the stars on evolution so abundantly manifest as in the path of humans from hunter-gatherers to farmers, technologists and space-farers. Beings that were denied the opportunity to marvel at the glory of the night sky may have developed entirely different belief systems, and hence cultures, from those that we humans have. This is just one example of how the cultural paths of two intelligent technological civilisations could diverge. Doubtless, there may be many more. We cannot as yet be certain if extraterrestrials share our curiosity for and affiliation with the wider universe. But it seems to me that in a galaxy of 100 billion stars, there ought to be many creatures who have had the immense good fortune to muse and marvel at the stars.

The stars will beckon to these creatures as strongly as they have beckoned to us. Some will develop technology that will allow them to traverse the great expanse of space between the stars. As we speak, vast amounts of information from an exotic alien civilisation may be making its way through the great void of interstellar space. We might have to wait many decades or even centuries to find it, but I believe that day will dawn on the human species.

What a joy it would be to know the culture, myth, science, history, literature, music, religion and philosophy of another civilisation in the Galaxy. Think how much the existence of extraterrestrial intelligence could enrich our outlook of the universe and our place within it. Like a lost child, our species yearns to find companionship, understanding and encouragement from other sentient creatures in the cosmos. In the turbulent times of our cosmic puberty, we are just beginning to illuminate this, the most fantastic of cosmological questions.

Chapter IV

And Death shall be no more, Death thou shalt die!

Like as the waves make towards the pebbled shore,
So do our minutes hasten to their end;
Each changing place with that which goes before,
In sequent toil all forwards do contend.
Nativity, once in the main of light,
Crawls to maturity, wherewith being crown'd,
Crooked eclipses 'gainst his glory fight,
And Time, that gave, doth now his gift confound.
Time doth transfix the flourish set on youth
And delves the parallels in beauty's brow,
Feeds on the rarities of nature's truth,
And nothing stands but for his scythe to mow:
And yet, to times in hope my verse shall stand,
Praising thy worth, despite his cruel hand.

Sonnet 60
William Shakespeare (1564-1616)

What are the hopes of man? Old Egypt's King
Cheops erected the first pyramid
And largest thinking it was just the thing
To keep his memory whole, and mummy hid;
But somebody or other rummaging
Burglariously broke his coffin lid.
Let not a monument give you or me hopes,
Since not a pinch of dust remains of Cheops.

Verse from 'Growing Old'
Lord Byron (1788-1824)

The immortal words prefacing this chapter eloquently express humanity's preoccupation with youth. The passage of time is often seen as the burglar of life or the leveller of Man's greed. The search for the elixir of life became an enterprise fraught with mystery and pseudoscience at its height in medieval Europe. Humans, in persistent denial of their mortality, hide behind a veil of religious obscurity, but deep down it stems from a basic fear of the unknown terminality of death. Would it be good to live longer than our biological 'use-by date' says we can? Is death as natural as eating and sleeping? Why do we age? Is there in any sense an elixir of life that will 'trick' death into believing our time has not yet come? In this chapter we shall explore these questions in the light of the increasing attention the subject of ageing is receiving from the scientific community.

Alchemy and Elixirs

The prospect of dramatically extending the human life span must have sparked the imagination of many people from the dawn of civilisation. The inexorable march of a human through childhood and adulthood and then through decline and old age must have been a powerful factor in the consolidation of our religious beliefs. The fact that one minute a person can be alive and well and the next minute dead, mute and cold, is sometimes too much for us to accept. Humankind may have invoked a deity to explain this transition. The death of a person could be seen as marking the boundary between the mortal world of humanity and the world of spirits. Death supposedly allowed the transmigration of the soul to its eternal resting place in heaven or, in the case of reincarnation, another bodily form. These religious beliefs, as we have seen, have been with the genus *Homo* for at least 200,000 years.

With the rise of science, it became possible to consistently use remedies from nature's pharmacopoeia to ease the suffering of sick or diseased individuals. Slowly and cautiously, humankind became convinced that the study of nature and her secrets would lead to an elixir that would ward off the descent into old age. The birthplace of alchemy is thought to be ancient Egypt, where in Alexandria it flourished under Greek influence. It was here, in the 5th century BC, that the Greek alchemist Empedocles of Acragas proposed that all things are composed of water, air, fire and earth, a notion upheld by learned individuals for over 1,500 years. Influential alchemical schools also emerged in China and Arabia. The masterly alchemical texts of the 14th-century Arabian alchemist Geber, whose earliest known work is the *Summa Perfectionis* (Summit of Perfection), describe all the major alchemical knowledge known to the ancients. Slowly, the zest for alchemy spread to Europe during the 11th and 12th centuries AD. Among the most influential of the alchemists at this time was the scholar Roger Bacon (1214-92), who believed that gold dissolved in a mixture of nitric and sulphuric acids (aqua regia) was the elixir of life. In reality, however, this is a highly vitriolic toxin, capable of killing anything remotely biological.

At the heart of alchemical philosophy was the belief that one substance could be changed into another by adding or subtracting elements in the right proportions. And although we now know that this cannot be achieved except in nuclear reactions, alchemists nonetheless made many important discoveries. They were the first to produce nitric, hydrochloric and sulphuric acids – substances underpinning industrial chemistry today. Alchemists also discovered the element phosphorus, from bizarre investigations of the properties of urine! The alchemist worked in a quasi-systematic manner to find a cocktail that would transform the serendipitous individual into an immortal figure, unbeguiled by mortality.

The records of alchemical practices are awash with peculiar pseudo-scientific recipes and unusual diagrams illustrating equally strange contraptions. Unscrupulous members of the alchemical community deliberately faked the production of gold from 'lesser' elements to ensure the continued supply of precious funds from their patrons, in order to sustain their researches. The practice of this somewhat occult discipline did a lot for science. Through alchemy, people began to realise that formidable truths about the nature of the world could be revealed by experiment and careful documentation of the observed events. The taste for experimentation – getting one's hands dirty – was to be become consonant with the philosophies of all future experimental sciences.

Many claimed to have found the elixir of life. All were frauds. The first mind to break free from the allure of ancient alchemical knowledge was the Swiss alchemist Theophrastus Bombastus von Hohenheim, otherwise known as Paracelsus (1493-1541). His extensive investigations led him to formulate the principles of the modern science of pharmacology. All physical afflictions, he asserted, resulted from chemical imbalances in the body. Paracelsus believed in the existence of an undiscovered element that he referred to as *alkahest*, from which were derived all other forms of matter. He also held that the isolation of alkahest from conventional forms of matter would provide the much sought-after elixir of life. The great mathematician and scientist Isaac Newton, better known for his epochal work on the nature of gravitation and the laws of motion, actually spent considerably more time on his alchemical researches than any other activity, including mathematics and physics. The great Italian theologian St. Thomas Aquinas (1225-74) was also attracted to the occult-like activities of alchemy, and widely advocated its use in the search for an elixir of life. These are clear reminders that intellectual brilliance does not guarantee foresight beyond contemporary ideas and activities. Yet it was the alchemists, now mostly unknown and forgotten individuals, who fuelled the modern search for something akin to an elixir to buttress the rampage of youth into old age. If anything, the work of these past experimentalists presages our wistful hopes for a longer, healthier life.

The Mechanics of Ageing

Ageing can be considered as a deteriorative disorder in which the organs of the body do not function optimally. Ageing is characterised by a slow but apparently inexorable decline in the metabolic, physiological and neuro-

logical functionings of the mammalian body. The key to ageing is undoubtedly one or more molecular phenomena. Let us try now to retrace,
as science sees them, the events leading to the demise of the beautiful
human form as it slips slowly into the grips of old age.

As we have seen in Chapter II, cells of the human body, except the
sex cells – the male sperm and the female egg cells – have 23 pairs of
chromosomes. These thread-like structures contain all the genetic information necessary for the manifestation of the human form. Sex cells
only have one copy of the 23 chromosomes and need another copy to
undergo cellular division. This is achieved through fertilisation, where the
sperm cell attaches to specific receptors on the outside of the egg cell
and penetrates it in an intricate courtship that lasts several minutes. The
nuclei, containing the genetic blueprints for a human being, are then fused
to form a new cell nucleus containing 23 pairs of chromosomes. It is only
after this event has occurred that this *hybrid* cell can divide and so begin
the formation of a new individual. Through successive divisions, the cell
produces more and more cells, first two, then four, eight and so on. Only
then does something dramatic occur. At the 16-cell stage a process called
differentiation occurs, where cells begin to sense somehow that they
belong to a developing brain or heart, liver or eye. Cells begin to migrate
to different regions of the developing animal, directed and orchestrated
by biochemical signals that tell the cell what to do next.

Cell division is apparent to all of us. We comment on the alarming
growth of children in the first few years after birth, and again at puberty
when the body undergoes a dramatic spurt marking the transition from
youth to adulthood. The body grows because more cells are born than
can possibly die. This continues all the way through childhood and adolescence and then a balance is reached which lasts for a decade or so
before signs of decline become apparent. When the body is fully grown,
the number of cells that are alive and dividing will equal the number that
die, so no net growth occurs. Instead, equilibrium is maintained and the
body goes into a steady-state period, which lasts throughout young adulthood and into middle age. Then, slowly but surely, the signs of ageing
begin to appear – going grey, balding and wrinkly skin are some of the
classic symptoms. As we get older, the number of cells that die off in any
period of time outnumber those actively dividing, and therefore, living.
This leads to a slow decline in the performance of the organs of the body,
and eventually they become so dysfunctional that death is inevitable.

Molecular biological evidence has shown that a battery of genes is
slowly inactivated as we age. Take the case of going grey. Deep down in
the hair follicles of your body there are cells called melanocytes, which
synthesise and secrete a dark-brown substance called melanin. The level
of secretion of this molecule determines the colouring of our hair. As we

age, however, more and more of our melanocytes shut down, ceasing to produce melanin and resulting in a gradual loss of hair colour. Changes in the pattern of hair growth are less well defined. Generally, as we age, the hair thins as the number of viable hair follicles decreases. However, hair may be stimulated to grow in other, more unusual parts of the body. For example, males tend to grow hair inside their ears or in the nostrils, while some females may develop facial hair, centred on the chin.

The reproductive capacity of humans also declines as we age. In males, ageing is characterised by a decreased ability to produce an erection, as well as a reduction in viable sperm cells. However, while male ageing is characterised by a reduction in reproductive capacity, females are fated to experience a complete cessation of fertility. This usually occurs between the ages of 45 and 50, and is known as the *menopause*. The onset of the menopause is characterised by a reduction in the length of the menstrual cycle from, on average, between 28 and 30 days at the age of 30, to less than 25 days at age 40. The steroid hormone, *oestrogen*, has a pivotal role to play in overseeing the development of the secondary sexual characteristics of females. After the menopause, the secretion of oestrogen from ovarian tissue dwindles, and as a result the structural integrity of the breasts becomes undermined. The menopausal woman undergoes a dramatic transformation as fat and water migrate from the breasts, causing the mammary tissue to sag and wrinkle.

Even the familiar changes in the texture of our skin can be explained in terms of a process that involves the inactivation of a set of genes. For example, the most abundant protein in the human body is *collagen*. This is a rope-like molecule that provides much of the structural integrity and elastic properties in healthy skin, supple joints, strong bones and active muscle. *Elastin* is another abundant protein that complements the properties of collagen in maintaining body flexibility. As we age, the levels of these proteins decrease as a consequence of the gradual inactivation of their encoding genes. This results in compromised plasticity and an impaired ability to repair damaged skin.

The Miracle of Modern Medicine

In the Stone Age, the average life expectancy of a human being was 25 years. This increased to no more than 30 years for the vast majority of people living in relative poverty up to medieval times. In the 1890s, the average life span of the British urban male was 47 years. By the early 1990s, life expectancy had increased to an impressive 76 years. In terms of human biology, this represents a huge increase in life span. New medications, together with improvements in nutrition and sanitation, have raised the average life expectancy in developing nations of Africa and Asia from as low as 40 in 1960 to 65 in 1990. And if this were not impressive enough, it is expected that children that are born to western parents in the 1990s will survive between to 95 and 100 years.

A Glimpse of Immortality – the Lowly Cancer Cell

In order to begin to understand more about ageing we must look at its conjugate, immortality. How can we glimpse what it might be like to live forever? We need look no further than the lowly cancer cell, responsible for the deadly group of diseases that have plagued humanity since the dawn of our species. The cancer cell is a living unit that has gone awry. Such cells, when given adequate nutrient provisions, will continue to divide for ever. In a sense, cancer cells have overridden their built-in cell death program. Normal cells replicate only a certain number of times, depending on the tissue of origin and the age of the individual. For example, cells that have been isolated from very young animals have been shown to undergo considerably more replications than older cells from the same tissue of origin.

A normal cell will accumulate a number of faults as it carries out its many and varied day-to-day chores. Once a threshold level of damage has been reached the cell is instructed to die so that new, more proficient cells can take its place. This process is called *apoptosis* (pronounced 'apo-tosis') by cancer biologists. Cancer cells often accumulate mutations in those genes which encode proteins that regulate the growth and division of cells. Many of these so-called *oncogenes* are incorrectly synthesised and perform abnormally. As a result, the unfortunate cell is fooled into thinking it should continue to grow and divide, and it proliferates beyond control, invading neighbouring tissue. Eventually it spreads all over the body via the lymph or blood vessels and then invariably causes severe damage and death to the unfortunate individual. Immortality is clearly not a good thing in the case of the cancer cell! The way to counter cancer is to fool the body into letting the old (normal) cells die and replacing them with just enough new cells to maintain optimum body function. Although this has never been achieved in practice, there are very important clues which many scientists consider promising. One phenomenon appears to take centre-stage in many deteriorative diseases. It's called the oxidative stress hypothesis, and it is intimately linked with our body chemistry.

The Oxygen-Powered Machine

When you eat, say, a ham sandwich, the digestive tract breaks the food up into a nutritious paste of its molecular components. Many of these molecules – all manner of sugars, fats and proteins – are utilised as valuable energy sources by our body cells. The actual mechanism behind this wonderful process is extraordinarily complex, but suffice it to say that the body strips the energy-rich molecules of their electrons which are used to convert molecular oxygen into water in a process called *respiration*. In order for the electrons to reach oxygen, they must be transported along an electron chain, which involves passing them sequentially to a number of proteins and smaller molecules, ending with oxygen.

The electron transport chain is located in a subcellular compartment called a *mitochondrion* (see page 52). The architecture of the mitochondrion is such that it contains two membranes, an inner and an outer. The force that causes electrons to traverse this transport chain also causes protons to be ejected from one side of the inner mitochondrial membrane to the other. This in turn produces a build-up of protons – a *proton gradient*. Using a mechanism analogous to a water dam, these protons rush through a number of molecular channels that are located across the inner mitochondrial membrane. This stream of protons is so energetic it enables the cell to harness the energy in the form of *ATP* (*adenosine triphosphate*), the universal energy currency of terrestrial life forms. Oxygen can thus be seen as a molecule essential for life processes, a gift of nature. But, as we shall see, it was a not a benefaction granted without a price. To understand the moral of this contorted tale, we must return once again to an epoch in the Earth's history before oxygen was present in the atmosphere.

The Debut of a Molecular Monster

As we have seen, the Earth's early atmosphere was most likely devoid of molecular oxygen. It is thought to have been composed mostly of nitrogen, as it is today, but also with appreciable levels of carbon dioxide, methane and ammonia. Compared with our present atmosphere, the air of the early Earth was considerably less *oxidising*. Life on Earth evolved by adopting a biochemistry that utilised the constituents of the primordial air. The first cells could not use oxygen to break down food molecules, but instead developed biochemical pathways to harness the greatest possible amount of the energy available from electron-rich chemicals such as those released by deep-ocean hydrothermal vents today. Within every cell we still see the vestiges of this ancient epoch.

Extant living things have maintained pathways that harness a small yet significant proportion of the energy from food molecules. For example, a pathway called *glycolysis* has been found in almost every cell from humble bacteria to those making up elephants. Glycolysis literally means the splitting of sugar. In this primitive pathway, a basic fuel molecule such as the six-carbon sugar, glucose, is split into two molecules of the three-carbon sugar, pyruvic acid. This process takes place in ten key steps inside cells, and releases but a small fraction of the energy stored in the glucose molecule. Much more energy could be released if the three-carbon pyruvic acid molecule could somehow be further broken down into simpler, one-carbon molecules like carbon dioxide. Yet, for almost half the history of life on Earth, cells did not evolve a more energy-efficient biochemistry.

Although the first living things were thought to have been consumers of energy-rich molecules, there must have been a limit to the available resources in the oceans of our planet. We shall never know for sure the chronicle of events that shaped the first ecological niches on Earth. Computer models, based on our best understanding of the evolutionary process, suggest that early food consumers, as they reproduced and increased in numbers, must have faced an energy crisis. They exhausted their food supplies.

Some organisms alleviated the problem by adopting a predatory way of life, killing and eating other cells. Others adopted a very different strategy that had profound repercussions on the development of complex forms of life on Earth: they invented photosynthesis. Instead of looking to ready-made sources of energy, some organisms learned how to couple the energy of sunlight to the production of sugars. Photosynthetic organisms learned how to make their own food. But there was another spin-off to this biochemical development. As they harvested the rays from a star eight light minutes away, these primitive plants released oxygen gas into the atmosphere – a blessing and a curse, as we shall see.

The Oxygen Holocaust

We normally consider oxygen as the bestower of life, the vital gas needed for our continued existence. This may be true today, but oxygen is in fact a highly reactive molecule. It is greedy for electrons and is capable of pulling them off most organic molecules. As a result, it can cause severe damage to living cells if it is not dealt with in an appropriate manner.

Widespread photosynthetic biochemistry was initiated well before 2 billion years ago. Although oxygen is highly toxic, it did not unduly perturb early oceanic life. Oxygen's low solubility in water ensured that living cells did not have to contend too much with its poisonous nature. Cells at first developed strategies to 'mop up' oxygen, restraining it from inhibiting various metabolic processes, but this could only work temporarily. The removal of oxygen was also facilitated by the chemical combination of oxygen with iron, becoming oxides in the Earth's oceans. But after several hundred million years, these biological and non-biological strategies to restrict the free availability of oxygen became saturated. Cells watched helplessly as the levels of free oxygen slowly climbed. Ill-equipped to deal with this gaseous onslaught, most forms of life perished in a so-called 'oxygen holocaust'. The evolution of molecular oxygen into the Earth's early atmosphere must have been a powerful natural selection event. Eventually, as a result of the pressure imposed upon replicating organisms, a cellular species evolved to use the toxic oxygen molecule in a radically new way – *respiration*. The cell figured out how to breathe. Oxygen came to be used as an electron acceptor to tap even greater amounts of energy from organic molecules. This heralded the birth of *aerobic organisms*, and it represented life's pacification of the oxygen molecule – at least for a while.

Oxygen – the Jekyll-and-Hyde Molecule

In order to form water in respiration, the oxygen molecule must receive four electrons which carry the reaction to completion. Sometimes, however, only one, two, or three electrons are picked up by molecular oxygen, producing highly reactive oxygen species, or *free radicals,* in the process. Let loose, free radicals have the potential to cause havoc with cellular machinery. Because the water molecule is far more stable than any oxygen free radical, the latter resorts to ripping electrons off nearby biomolecules in order to become a more stable water molecule. This initiates a chain reaction in which molecules making up the various components of the cell become radicals in their own right. Unless these reactions can be quenched somehow, these tenaciously active molecules will cause an irrevocable domino effect of damage to the cell.

You might wonder what all this can have to do with ageing and cancer. The connection is the considerable damage free radicals can exert on our genes. By reacting with the individual bases that constitute the core of DNA, radicals change DNA's chemistry and functioning. This leads to progressively less efficient replication of the DNA molecules, as well as the accumulation of mutations that may produce faulty enzymes, leading the cell astray. The results are cancer, ageing and eventually death. Although oxygen is necessary for the life function of all aerobic organisms, it also has its price. The production of free radicals is an unavoidable process. Through our continued respiration, our mitochondria continue to produce billions of these radicals and they continually diffuse out into the cell to prey on whatever tickles their fancy.

There is abundant evidence to suggest that free radicals are indeed associated with the ageing process. For example, if rats are placed in an atmosphere of hyperbaric oxygen, having a higher concentration of oxygen than is found in normal air, they rapidly develop damage to their lung tissue and age much faster than siblings raised in normal air. If left in this environment for only a few hours, they die. Furthermore, research with nematode worms shows that individuals age more rapidly if exposed to increased concentrations of oxygen.

From the moment of its birth, a cell respires continuously, taking in oxygen and using it to strip electrons from food molecules, and releasing carbon dioxide and water in the process. As the cell matures, it sustains more and more damage to its inventory of proteins, sugars and fats, as well as its

DNA. It has been estimated that approximately two to three per cent of the oxygen consumed by aerobic cells is diverted to the production of reactive oxygen species. Calculations have suggested that a typical mammalian cell may, in fact, suffer 100,000 radical attacks on its DNA per day. At any one time, almost 10% of our cell's proteins are oxidatively damaged. The basic idea is that as we age, we accrue more oxidative damage, leading to a steady decline in physiological performance, culminating in death. Plainly, oxygen is a molecule with two personalities. As the dissembling harbinger of both life and death, oxygen is a Jekyll-and-Hyde molecule.

Fortunately, it's not all doom and gloom! Human cells have had billions of years to prepare defences against the rampaging oxygen radicals. Let's take a closer look at some of them.

Our Gallant Protectors

The first line of defence our cells possess is a ready supply of so-called *antioxidant* molecules. The commonest ones include chemicals such as glutathione, isoflavonoids, Vitamin C and Vitamin E. These antioxidants serve to 'mop up' free radicals, quenching their ability to damage biomolecules. The second line of the cell's defence is a host of antioxidant enzymes that catalyse the conversion of aggressive free radicals into less tenacious molecules. For example, an oxygen molecule that receives only one electron is called a *superoxide*. Aerobic cells synthesise an enzyme called *superoxide dismutase* (*SOD*) which transforms this reactive molecule into hydrogen peroxide. In turn, the less reactive hydrogen peroxide gets degraded to water by an enzyme called *catalase*. As we age, however, our first and second lines of defence against reactive oxygen species become weaker. The levels of our small-molecule antioxidants plummet as we enter old age, and our bodies are not as well able to maintain the synthesis of adequate levels of SOD and catalase.

But is this really evidence that oxidative stress contributes to the ageing process? Indeed, is oxidative stress a cause or a consequence of ageing? An encouraging clue has arisen from genetic studies of fruit flies. A group of researchers determined the effects of deliberately overshooting the levels of catalase and SOD in the fruit fly *Drosophila*. The results were astounding. Flies that were genetically engineered in this way displayed an average 33% increase in maximum life span over normal flies, a pronounced retardation in the age-related accrual of oxidative damage to DNA and a significant decrease in the rate of mitochondrial hydrogen peroxide production. Together, these observations strongly suggest that the alleviation of oxidative stress could help to increase the life span of aerobic organisms, including humans.

Bearing these results in mind, one might then expect the life expectancy of animals with a high metabolic rate, that is, highly active creatures, to be considerably lower than that of similar-sized animals with a relatively low metabolic rate. In fact, in nearly all situations this prediction turns out to be true. Take the case of two creatures, equally small (or large), who have well-characterised but radically different metabolic rates. There are such comparisons available to us – the weasel and the sea turtle, for example. Pound for pound, there is little difference in these creatures, but because the turtle has a slower metabolic rate than the sprightly weasel, its longevity is more than one order of magnitude greater. One thing appears to be clear: if we want to live longer we must counter-act the production of free radicals. How can this be achieved?

Several decades ago a number of scientists advocated the restriction of calorie intake as a means to extend longevity. Their reasoning was simple. They would place the individual on a diet with considerably fewer calories for an indefinite period. With a reduced daily intake of food, the individual would, in theory, burn fewer fuel molecules and therefore pro-duce fewer free radicals. Careful studies have shown that a number of animal species respond positively to calorie restriction. It is possible to extend the life span of organisms as divergent as flies, guppy fish, spiders and rodents simply by reducing the amount of calories in their diets. The extensions beyond their normal life expectancy vary between 15 and 35%. Preliminary experiments with primates suggest that similar exten-sions of life span are possible using calorie restriction.

Ageing and DNA

Free-radical damage is not the only agent of ageing, however. Over the last few decades the genetic contribution to the ageing process has be-come well documented. A flood of new information has been released by the application of the powerful methods of molecular biology. The first indication that genes, at least to some degree, dictate the rate at which we age, comes from studies of flies and worms. For example, it has been known for many years that it is possible to select individual strains of flies that live longer on average than other strains from the same species. Likewise, in the nematode worm *Caenorhabtidis elegans*, genetic selection has allowed for the isolation of a mutant, termed Age-1, which extends life span in these invertebrates by 33%. These longer-lived worms display sluggish behaviour and, curiously, a signifi-cantly increased tolerance to heat shock.

Heat-shock proteins are synthesised in response to a number of environmental stresses, including oxidative stress and, as their name sug-gests, heat shock. They represent a family of proteins that function to recoup structurally damaged proteins – 'molecular chaperones'. Careful research has shown that the levels of some heat-shock proteins synthesised within our cells decline as we age, thereby further decreasing our capacity to repair damaged proteins. Furthermore, scientists have presented evidence indicating that the genes responsible for synthesising heat-shock proteins are inactivated after a threshold level of ageing has been reached.

These observations suggest that increasing the levels of heat-shock proteins in our cells could help retard the onset of cellular senescence. It is important to emphasise that the Age-1 mutant worms do not have an extended old age in the normal sense of the word but have a capacity to maintain their body chemistry longer and more effectively than normal or 'wild type' worms. The Age-1 mutant is not the result of a single gene change, but of a number of genetic factors working co-operatively to slow down the ageing process.

Using similar genetic approaches, researchers have discovered and partially characterised a variety of genes, from other organisms, that encode factors contributing to the ageing process. What is very surprising is that several unrelated genes with completely different functions are capable of conferring increased longevity on the organism. Some of these genes encode proteins that function as receptors for steroid and thyroid hormones, others code for growth factors, and still others are involved in the replication of DNA. It is apparent that the genetic factors that help determine the rate at which an individual ages are many and varied. As with the origin of our intelligence, there may be a strong genetic disposition to ageing dictated by a number of genes acting in concert.

Advances in genetic research have also identified enzymes that help to replicate our DNA. As we have seen, all advanced cells, including those that make up the human body, are composed of a number of long, slender, thread-like structures called chromosomes. During a cell's life, chromosomal DNA undergoes a number of replications. However, there appears to be an upper limit to the number of divisions a human cell can undergo in the course of its lifetime. This is called *replicative senescence*. Estimates put this number at or near 50 divisions. Moreover, the cell retains a strong memory of the number of divisions it has already undergone. For example, laboratory-grown cells that have divided 10 times, but are frozen for many years and then slowly thawed, appear to undergo no more than 40 further replications. The cell's replicative memory is formidable. It is not fooled by many years of frigid dormancy.

As I have mentioned, the capacity a cell has to duplicate is related to its age. For example, cells that are isolated from very young animals can replicate many more times than similar cells isolated from an aged member of the same animal species. For whatever reason, older cells cannot replicate as well or as many times as their younger counterparts. If we examine the fine structure of a chromosome, we find regions that are necessary for initiating the cell division cycle. For example, the *centromere* is, as its name suggests, located centrally along the chromosome and serves to bind the entire chromosomal structure to the *spindle fibres*, thread-like appendages that help orchestrate the replication and eventual division of the chromosome into two *daughter chromatids*. Located far out on the chromosome tips are the *telomeres*, structures that contain highly repetitive DNA sequences – 'junk' DNA.

Telomeric sequences contain no real genetic information and thus do not encode useful proteins. But telomeres were shown in the early 1970s to become curiously shortened as cells underwent successive rounds of replication. Telomeric sequences are not very long, and because they play an integral role in cell division, they eventually become too short for replication to proceed efficiently. When this happens, the cells cannot divide and the result is apoptosis. We can picture telomeres as molecular clocks counting down the remaining cellular time. As cells go through successive divisions, that is, as they age, their telomeres get progressively shorter. As a result, we can use telomeric length in any given cell type as an indicator of relative senescence. These observations provide a powerful, rational explanation for cellular ageing – or do they?

Research on several cancer cell lines has revealed that although most have shortened telomeres corresponding to the number of replications undergone by the cells, there are a significant number of cancers where telomeric length does not correlate with the number of cell replications undergone. Further work has revealed the presence of an enzyme called *telomerase*, which functions to add on new pieces of DNA to the shortening telomeric sequences. Some cancer cells were found to produce very high levels of telomerase, although this was certainly not the case for all tumour tissue studied. Nonetheless, biologists quickly realised that telomerase could be a candidate enzyme to help extend life span. As in the case of heat-shock proteins, by learning how to pump up the production of this enzyme in cells, scientists might be able to prevent replicative senescence without turning normal cells into immortal cancer cells.

Clearly, the story of telomerase provides us with some valuable lessons on the nature of ageing. It is likely that there are many pathways leading to cell death and that many more events in cellular senescence remain to be discovered. Although we are certain that genes play a pivotal role in the development of cellular ageing, we are not sure of the consequences of tampering with genes like telomerase or heat-shock proteins. Normal cells are designed for death, and it is only when their programming is bypassed that we see the capricious tumour cell emerge.

Sex and Mortality

Cells vary enormously in their capacity to replicate. For example, most bacterial cells, when supplied with optimal energy and mineral resources, will continue to divide every 20-30 minutes. If this nutrient supply remains constant, such cells can be considered immortal. Bacteria and almost all other forms of single-celled organism divide by *binary fission*. This is a form of asexual reproduction simply involving the replication of its DNA genome, followed by a co-ordinated segregation of the cellular components into two smaller *daughter cells*, each genetically identical. Bacteria, it seems, produce twins all the time. Multi-cellular organisms, by contrast, cannot normally replicate by simple binary fission, but instead tend to reproduce sexually. Thus, parental organisms live on, overseeing the development of new siblings. Sexual reproduction was selected in the winding course of evolution because it allowed a more efficient 'mixing' of genetic traits. During fertilisation, DNA pieces from male and female sex cells complement each other, patching over the genetic weaknesses of both organisms. The considerably greater genetic diversity achieved through sexual reproduction is to a large degree responsible for the outstanding success of multi-cellular organisms, in terms of adaptability and evolutionary design. The price we pay for having adopted sexual reproduction into our life cycle is ageing and death.

Sleeping Beauties

Some mammals, such as squirrels and brown bears, enter a deep sleep during winter. A highly adapted hibernator, such as a ground squirrel, retires underground in the appropriate season and reduces its body temperature and metabolic rate. During hibernation the animal's metabolic rate may be decreased by 10-15% and its rate of heartbeat can decrease from an average of 200-300 beats per minute to a sluggish 10-20 during hibernation. An increase in air temperature signals the arrival of spring and causes the hibernating animal to produce a pulse of increased metabolic activity, completely restoring its normal physiological pattern.

Biologists have studied the complex physiological changes that occur in hibernating animals. Research suggests that substantial changes in the levels of circulating hormones, such as thyroxin and other regulators of basal metabolic rate, may be responsible for inducing hibernation in mammals. Although the precise biochemical events that trigger hibernation have not yet been elucidated, there is a distinct possibility that a hibernation-like state might also be achieved in humans.

When we sleep, arterial blood pressure falls, pulse rate decreases, most body muscles relax and the overall basal metabolic rate decreases by about 10-30%. Sleeping causes the decrease in basal metabolic rate, significantly reducing the rate of free-radical attack on cellular components, and allowing the body's repair machinery to work more effectively. This explains why sleeping has for so long been associated with renewal, repair and revitalisation of the body. Hibernation may thus be viewed as an extreme form of sleep in which basal metabolic rate is decreased even more and many vital functions become dormant. As most people know, we sleep less as we age. This may indicate that the processes that operate to rejuvenate the body during sleep decline in proficiency as we age.

By reducing basal metabolic rate below the level attained by sleeping, it may be possible to induce a hibernation-like state in humans. This would reduce energy expenditure, thereby decreasing the output of reactive oxygen species and increasing the average life span. Just how far science can push this reduction in metabolism is a matter of conjecture. Too little will yield results no better than those obtained by normal sleeping; too much, and the unfortunate individual could end up with extreme somnolence or even in a permanent coma! A hibernatory state of animation in humans may be of use in the exploration of space, particularly if technologies capable of attaining a large fraction of the speed of light are not feasible. For example, a state of suspended animation would allow humans to travel a distance of ten light years and back, at a speed of ten per cent of the speed of light, by completing the mission in approximately 200 years.

Closer to home, it may be possible one day to administer drugs that will significantly reduce basal metabolic rate only during sleeping periods. This may have the effect of increasing the efficiency of cellular repair, reducing the oxidative burden on the body and, over a period of many years, increasing average life span significantly. If this ever comes to fruition, then we may not have to work very hard at all to enhance our individual longevity.

The Murder of Mind – Alzheimer's Disease

For many individuals, ageing is also intimately associated with the loss of mental acuity. Dementia in some form or other seems to come to many individuals in the autumn of their life. The statistics speak for themselves: five per cent of all individuals at age 65 and 20% of people over age 80 suffer from *Alzheimer's disease (AD)*. With the exception of heart disease and cancer, AD kills more people in the western world than any other disorder.

In 1908, the German physician Alois Alzheimer (1864-1915) reported a striking difference in the brain sections from normal individuals and from those who had succumbed to a common type of senile dementia. When he used special stains to improve contrast in brain tissue, his microscope revealed a number of abnormal-looking plaques and tangles that were conspicuously absent from normal brain tissue. In honour of the important discovery, this common senile dementia was named after him. The identification of the senile plaques and tangles still remains one of the most reliable methods of diagnosing AD, albeit from post-mortem tissue.

Alzheimer's disease is characterised by a progressive senile dementia, with sufferers experiencing a profound loss of memory, orientation and judgement. In the later stages, the person enters a 'twilight world' with loss of consciousness and finally death. There is, as yet, no cure for this debilitating disease. Our society seems to have accepted as inevitable the decline of the higher cognitive functions associated with the brain as we age. Going a bit 'batty' is commonly regarded as part and parcel of getting old. In many ways our complacency and ignorance about dementia is ingrained in our cultural heritage to such an extent that many refuse to accept that conditions such as AD are really illnesses, just as worthy of investigation and prevention as cancer or cot death. Here, too, we shall see how human courage and convictions have made considerable progress in understanding this deadly decline, with age, of our intellectual abilities. Through science, our changing attitudes to AD have yielded many exciting insights into the true character of senile dementia.

AD is broadly classified into two types: the so-called *early onset* AD, which afflicts individuals as young as 40 or 50 years of age, and the far more prevalent *late onset* AD, which manifests itself in individuals over the age of 65. The incidence of AD in females is greater than in males, which may simply concur with the greater life expectancy of women.

Ever since Alzheimer's discovery in the early twentieth century, clues about the nature of AD have been collated from a variety of studies. For example, meticulous biochemical analysis of brain tissue from AD patients revealed a general deficit of important chemicals, called *neuro-transmitters*, which function to activate or inhibit nerve cells in relaying myriad electrochemical signals. These neurotransmitters include *serotonin*, an important chemical that dictates mood and behaviour in humans and other mammals, and acetylcholine, an abundant molecule in both the central and peripheral nervous systems that plays a pivotal role in the orchestration of memory and motor co-ordination.

Some scientists reported an increase in the levels of aluminium salts within the senile plaques, leading them to speculate that a slow build-up of this toxic metal in the brain was the cause of the AD dementia. Further studies tried to demonstrate a causal association between high concentrations of aluminium in drinking water and an increased incidence of AD. The results were ambiguous at best, except for one clear-cut case – *renal dialysis encephalopathy*. This is a type of dementia that arose in patients who were forced by kidney failure to undergo regular dialysis using water containing appreciable concentrations of aluminium. Over a lifetime, millions of litres of water were used to clean their blood of its impurities and as a consequence the levels of aluminium and other metals slowly built up in the brains of these unfortunate individuals.

Most of the patients who underwent dialysis with this aluminium-enriched water developed a slow, debilitating dementia similar in many respects to AD. When the connection with aluminium was established, some leading physicians advocated the removal of all aluminium cooking utensils from the market to halt what they perceived as a slow, mass-intoxication of the population. Chemists responded by ridiculing these recommendations, claiming that barely soluble aluminium could not attain a sufficiently high concentration for such intoxication to occur. The argument continues. Today, however, due largely to the sporadic appearances of this type of dementia, the aluminium hypothesis has taken a back seat to more attractive theories, revealed through the powerful sciences of genetics and molecular biology.

Plaques and Tangles

The first clue to the genetic nature of AD came from series of studies demonstrating that some forms of this dementia were inherited. Some families were afflicted with AD more than others. The discovery that *Down's syndrome* patients, who inherit three copies of chromosome 21, developed lesions in their brain tissue which were similar if not identical to those in AD patients, added new impetus to the search for the ultimate cause of AD. It became clear that one or more of the genes located on chromosome 21 led to the production of abnormally high levels of proteins associated with the senile plaques and tangles in the AD brain. The isolation and subsequent characterisation of the proteins from senile plaques revealed that they were composed almost entirely of a protein called *amyloid.*

Further genetic work revealed the location of the gene encoding the amyloid protein. To the great excitement of many researchers in the field, the gene was localised to chromosome 21. Analysis of the structure of the amyloid gene demonstrated that there were important differences between the plaque protein and the protein encoded by the gene: it appeared that the protein was synthesised as a larger, precursor protein. The so-called *amyloid precursor protein (APP)* would have to have been subsequently modified by clipping off particular amino acids in order to produce the smaller amyloid plaque protein. Such a phenomenon is called *proteolytic processing* and is commonly observed in the production of digestive enzymes. For example, our intestinal cells synthesise and secrete, among others, the enzymes *trypsin* and *chymotrypsin*. These enzymes catalyse the breakdown of large food molecules into smaller, more readily utilisable substances.

In their active form, proteolytic enzymes will digest the interior of the cells from which they are derived. But nature has come up with an ingenious method to prevent this *autodigestion* from occurring. The cell produces the enzyme in an immature or *precursor* form. Such *proenzymes* are secreted from the cells lining the walls of the pancreas and intestinal wall and only then do they undergo processing to the fully active enzyme. Such proteolytic clipping can transform a biochemically docile protein into a fully active enzyme. In this way, the cell can avoid being eaten from the inside out.

The amyloid plaques found in AD brain tissue are also the products of this type of biochemical processing. The APP is ordinarily embedded in nerve cell membranes and appears to have some role in signalling between cells. The modified protein in plaques seems to have a greater tendency to aggregate into insoluble plaques that disrupt normal neuron functions, such as their ability to communicate with other nerve cells nearby. So AD could be triggered in people who are predisposed to processing the APP in an abnormal fashion. We still do not know whether this amyloid deposition is really a cause or merely a consequence of the underlying pathology of AD, but all the indications are that amyloid deposition in nerve tissue could severely compromise the functioning of our brain.

Oxygen Again!

Some scientists have looked for other differences between AD brain tissue and ordinary brain tissue. As well as a general deficiency of common neurotransmitter molecules needed for proper nerve-cell signalling, there is a dramatic degree of brain tissue damage caused by oxidative stress. By examining thin sections of post-mortem tissue, scientists have found that extensive regions of the AD brain have apparently been damaged by oxygen free radicals. Some researchers attest that this oxidative damage may cause or at least contribute to the massive atrophy of the cerebral cortex in AD patients. Moreover, it is well known that the production and propagation of reactive oxygen species may activate the inflammatory response. It has been observed that rheumatoid arthritis patients, who are regularly administered anti-inflammatory drugs, have a significantly reduced risk of developing AD. Could these drugs prevent or ameliorate the widespread inflammation seen in the AD brain? And would anti-inflammatory drugs act to reduce the production of radicals and the damage to ageing nerve tissue?

Preventive Measures?

It seems clear that the generation of reactive oxygen species and the cumulative damage incurred by their production provide a rational explanation for many degenerative diseases. Theories of free-radical damage to tissue have been invoked to explain a whole host of other degenerative diseases including multiple sclerosis, motor neurone cell disease, rheumatoid arthritis, atherosclerosis, many forms of cancer and Parkinson's disease. With this knowledge, is there anything we can do now to prevent or at least delay the approach of physical and intellectual decline? I believe there is.

We may be able to develop drugs that boost the levels of our antioxidant defences. This would involve 'kick-starting' inactivated genes such as catalase, superoxide dismutase and the genes responsible for the biosynthesis of glutathione and heat-shock proteins. Similarly, some pharmaceutical companies are already seeking to produce new drugs that could prevent the deposition of amyloid protein. Others are looking for ways to interfere with the processing of APP into the plaque-forming amyloid protein. If indeed the amyloid depositions are the cause of the underlying pathology of AD, then these drugs of the future offer real hope of counteracting or even preventing the onset of AD. Further research will undoubtedly provide detailed information on the precise battery of housekeeping genes that are progressively inactivated with advancing age. It is conceivable that therapeutic agents may be developed to activate reproductive, structural and antioxidant genes in concert, providing a more thorough, well-rounded defence against ageing.

For the moment, we can supplement our diets with more nutrients that inactivate free radicals. For example, Vitamin C, Vitamin E, beta-carotene and a variety of isoflavonoids found in fruits and herbal beverages have a considerable antioxidant effect. It is unlikely that the maximum benefit can be achieved by taking these antioxidants by themselves. The greatest effect is more likely to be attained by eating fresh foods that are rich in these nutrients. Thus the inclusion of more citrus fruit in our diets will provide a ready source of Vitamin C; unprocessed wheat germ and oily sea food are rich sources of Vitamin E. The ingestion of whole foods rather than the active ingredients alone is of paramount importance because other, as yet ill-defined factors in the food may be necessary for the efficient uptake of these vitamins.

A Delicate Balancing Act

We have seen that molecular oxygen can be our friend or our foe. Aerobic organisms that evolved the capability of harnessing the oxidising power of molecular oxygen to extract energy from food molecules more efficiently paid a great price. Our biochemistry has been formulated to live with both faces of the oxygen molecule.

But before we pronounce our sentence on oxygen, it is also important to realise that some cells actually use oxygen free radicals to their advantage. Take the case of the humble *neutrophil*, a white blood cell whose role is to engulf invading bacterial cells. Once inside the cell, in *food vacuoles*, the neutrophil undergoes a dramatic oxidative burst, which not only synthesises reactive oxygen species but also directs these radicals to the food vacuole. The engulfed bacterial cell is severely damaged by this radical onslaught. Neutrophils can also secrete billions of molecules of reactive oxygen species in order to kill bacteria located outside their cells. These examples serve to illustrate that our bodies have clearly created a pact with the capricious radicals in order to maintain the delicate balance of life.

Knowing all this, what are the chances of finding an end to humanity's mortality? It can be argued that the maximum practicable life span for a human being is about 150 years. Our bodies are machines in most senses of the word, and our inventory of organs are subject to mechanical decline and eventual failure. The mammalian heart has an upper limit to the number of times it can beat, dictated, at least in part, by the ability of heart muscle to keep contracting and the frictional forces that tend to grind the heart to a halt. Put another way, because we are mechanical entities, we inevitably meet with mechanical failure. The short-term solution would be to remove an aged organ and replace it with an artificial or genetically engineered organ. This process of cybernetic part-replacement could, in theory, go on indefinitely. But the brain is also an organ, so it would have to be replaced at some stage during an individual's existence. It is here that we are confronted with a profoundly difficult predicament. How do we replace a brain? More importantly, how could we begin to restore the wealth of neural circuitry that had come to symbolise that individual's mind and memories?

Unless we can find a way to perfectly preserve the vast repository of knowledge stored away in our brains, we would have no way of perpetuating the individual, the self, within each of us. For the moment, it is clear that oxidative stress and its alleviation could provide a mechanism to extend our life span by at least 20-30%. The use of antioxidants and anti-inflammatory drugs to keep inflammatory processes at bay in the brain might well prove sufficient to prevent the onset of dementia. This is half the battle won – our minds would be healthy. To maintain a healthy body from the neck down requires a careful balance of good nutrition and adequate, but not excessive, exercise. It may be possible to design drugs that can mobilise our antioxidant defences more effectively by stimulating their synthesis or retarding their degradation.

The Technological Methuselah

Whatever the methods by which we extend the average human lifetime, the repercussions on our society may well be enormous. Our burgeoning human world population is putting the squeeze on government health-care policies. Our homes for the elderly are filling up fast. In Japan, for example, only seven per cent of the population was over 65 years of age in 1970. This had doubled to more than 14% in 1995. Such growth in the elderly population is typical of many developed nations, and although steps are being taken to provide adequate care and provision for the ageing community, it nonetheless poses problems in the global arena. Similar trends in developing nations are imposing additional burdens on health-care facilities that are already stretched to the limit.

As a mammalian species, we already have a reasonably long lifetime. Our culture is intimately linked to our biological longevity. If we lived longer we would have to restructure our society completely. Provision would have to be made for the dramatic increase in elderly human populations in the world. If we do find all the genes that predispose us to age, then their manipulation may cause *Homo sapiens* to have an extended, stabilised middle age. Future governments would have to radically rethink how to accommodate people in occupations that may last eight decades or more. Perhaps in such a society a human being could look forward to three or four full careers.

But with the ever-increasing global human population, it is clear that any immediate measure to introduce extended longevity to all members of society would have disastrous consequences. For now, the extension of life span is likely to remain a luxury for the rich. Regardless of whether or not our descendants can sustain a society of Methuselahs, it is clear that our species has embraced the idea of enjoying an extended longevity. It is yet another great yearning we have to conquer nature, to create an unnatural world. Now in our cosmic puberty, we realise that our mortal desires to live longer do not even come close to the ultimate aspiration of our species – immortality.

Chapter V

Returning to the Centre of Meaning

Finding myself to exist in the world, I believe I shall, in some shape
or other, always exist.   Benjamin Franklin (1706-90)

The waking have one world in common, but the sleeping turn aside
each into a world of his own.   Heraclitus (?535-?475 BC)

Nothing ever becomes real until it is experienced.
                                            John Keats (1795-1821)

In terms of size, we humans lie about mid-way between an atom and a
star. Yet with our powerful brains we have come to figure out many of
the secrets of the universe. The human imagination has conceptualised
tiny subatomic particles and walls of galaxies extending over hundreds of
millions of light years with equal facility. Our soaring intelligence and
curiosity have enabled us to discover the subtleties of nature, and to distil
truths that lie behind its elusive cast.

Science has revealed a universe of astonishing beauty, size and age.
Yet, in a cosmos with hundreds of billions of galaxies each containing
billions of stars, our perspective seems petty and insignificant. Many nearby
stars have been shown to possess planetary families of their own. Out
there, circling other suns lie a vast menagerie of planets – some like Jupiter
and Saturn, some like Mars and the Moon. But what we yearn for most
are other Earths, replete with cool water oceans and blue nitrogen skies.
We seek worlds that have spawned life and intelligence. We have not yet
found this life, but science gives us strong reasons to believe that it exists.

We have seen how, after 4 billion years of biological evolution, a
remarkable creature evolved to dominate the food chain. With a massive
brain, these creatures made the transition from insentience to sentience.
From our fully arboreal ancestry, our species slowly learned to adapt to
the changing environment of the eastern African continent. We evolved
an upright posture that allowed us to stay cool and carry valuable materials,
including weapons to hunt with and food to eat and share with our loved
ones. Changes in the dexterity of our hands allowed our ancestors to
manipulate objects to an extraordinary degree, inducing in us our latent

talent for making tools. And with them, humanity has worked miracles.

From simple stone chippers to quantum lasers, our species has blossomed in leaps and bounds. The emergence of human consciousness was the great liberation of our species, allowing us to understand and predict the vagaries of our planetary home. But we have also begun to use these tools to figure out how we got here, and where we came from. By putting our trust in science our species has discovered the blueprint of life. Through the molecular verse of DNA, the song of life can be sung. But we humans have the potential to remove lines from this ancient verse and replace them with others. We can change ourselves for better or for worse.

Modern medicine has raised the quality of human life immeasurably. Through better sanitation, new drugs and proper nutrition, humans have already outsold their biological 'use-by date' by nearly 300% in just a few centuries. This astonishing feat has awakened in us a yearning for greater longevity and, for some, even immortality. Perhaps in a century or two, we may have a stabilised society of Methuselahs equipped with the technology to voyage to the nearby stars. This great dream has captured the imagination of our contemporary society, for we are the first generation in the long and tortuous history of our species to begin the homesteading of other worlds. In the early twenty-first century there will be human-occupied bases on the Moon and on Mars. But our aspirations carry us farther than the Red Planet. Steady technological developments and perhaps even a few scientific and political paradigm shifts will be necessary before we seriously contemplate journeys to the distant stars. We have both the desire and singular ability to bring this great dream to fruition.

The vastness of space tells us clearly that our biological perspective may not be unique. The cataloguing of worlds orbiting other stars in the Galaxy and our search for extraterrestrial intelligence are well under way. If we are fortunate, contact will be established with other technological civilisations early in our future. Slowly, we may become interlocked in what the English writer Edward Ashpole has called an 'immortal network' – a majestic interconnection of a great many thinking beings, a union of knowledge-harvesters strewn along the spiral arms of the great Milky Way Galaxy. As we speak, a vast amount of information may be making its way across the cold dark of interstellar space. Someday, perhaps soon, we will detect it. Just as today's worldwide internet has revolutionised human commerce, education and entertainment, so too will the 'immortal network' cause a sea-change in our cosmic perspectives.

But our species has already made a bid for immortality – not just with anti-ageing creams and antioxidant wonder drugs, but through our radio transmissions. Since the introduction of radio technology in our society, our television and radio broadcasts have voyaged about 60 light years into space. Their lonely sojourns will carry them through clouds of gas and dust and to stars with barren, lifeless worlds, and on through shells of matter blown off in the death throes of dying stars and even to systems with life-bearing planets. We have every reason to believe that some of these worlds will be inhabited by creatures who, like us, have taken the step from insentience to sentience – beings who are separated from us not in time but in space. What a joy it would be to know something of the science, culture and philosophy of an exotic alien civilisation – to share our visions of the cosmos with other thinking creatures.

Our knowledge of nature, acquired at great cost over many millennia, has created a kind of destiny for our species. If we learn to survive the heady times of our technological adolescence, through the fostering of a greater degree of intellectual openness in research and education for all members of our society, as well as by moving towards economic and political unity, then our species has the glorious potential to take the next step towards a kind of cosmic consciousness. Just as the Earth's biosphere acts as a self-regulating unit, so too might a symbiosis of sentient creatures in the Galaxy create a greater conscious whole.

For now, the human perspective, set against the immensity of space and time, is like that of a single nerve cell in the deep interior of an immense brain. Only by interconnections between other nerve cells – other beings in the universe – can the semblance of cosmic consciousness be achieved. And like the brain, whose cellular components live and die, to be replaced by new cells, so too might our cosmic culture be maintained by the coming into existence of new societies and the dying off of older societies. But in coming together, in the establishment of a great congregation of conscious beings, our lives will be enriched immeasurably. Think of the knowledge and wisdom we will gain from the collected works of beings spanning aeons of cosmic time. It would change everything.

Science has well and truly removed our species from the centre of the physical universe. But although we are finite beings searching desperately for a wider context to our lives, we may find solace in what we are. A human being is a collection of atoms that have come together in an extraordinarily complex way to manifest the human mind. Yet the origin of consciousness is a profound mystery. On our little world, at least, it arose unexpectedly and without precedent. And while some scientists believe that it emerges after a certain level of complexity arises in the nervous system, others, myself included, cannot reconcile it with any natural process. For me, the emergence of sentience from insentient matter is miraculous. Its arrival in the world presents a sharp break in the natural order. What's more, consciousness is the seat of our moral agency – a behaviour that has no counterpart in the animal world.

With our minds we have come to extract beauty, symmetry, harmony and goodness from the universe. We seek a connection between the majesty of the cosmos and our fleeting existence as finite beings. And there is a connection. As conscious agents, as creatures who strive to weave meaning from a sea of indifference, we already enjoy a unique perspective.

In this book I have brought you across billions of light years of space. I have brought you deep inside the living cell with its myriad atoms and molecules. We have contemplated stars smaller than atoms and atoms larger than cities. We have peered into the heart of the human body to discover how the very air we breathe conspires to ordain the demise of our corporeal existence. But the source of all these questions lies with our consciousness, our passionate desire to know everything there is to know. We crave knowledge in all its forms, both rational and irrational.

Our deepest scientific questions about ourselves and our significance in this vast and imposing universe are fuelled by our innate beliefs. When theoretical physicists work their equations to find the ultimate law governing all of nature, they are impelled to do so by a *belief* that such knowledge is possible. When biologists strive to understand the ageing process and the origin of life, they do so from a *belief* that life and death can be understood by reason alone. When anxious radio astronomers search the heavens for signals of intelligent origin, they are driven in the *belief* that technological civilisations are scattered across the cosmos.

And even though we live in a society steeped in technological innovation, we humans show clear signs of expressing a kind of spiritual desire to connect with aspects of our being that are not demonstrably tangible – a part of our existence that science and its methods can never probe. The remarkable resurgence of interest in the paranormal, astrology and other forms of pseudoscience in the twentieth century are outpourings of our most deep-laid desires to connect with that which is beyond ourselves – to nourish a part of our being that science has made forlorn. Yet, remarkably, we use our science to stimulate our aesthetic sensibilities – a side to us that does not lend itself easily to scientific analysis. Through virtual reality, hallucinogenic drugs, fictional literature, music, art, theatre and cinema, we humans go to great lengths to remove ourselves from monotone reality. We are carbon-based aesthetes!

We realise that there is more to our existence than science gives us credit for. But while our most penetrating scientific questions have dethroned humanity from the epicentre of the physical universe, they have served to return us to the centre of its meaning and purpose. We are special. We do have a unique outlook. Only now, in the impetuous times of our cosmic puberty do we realise that we are the heirs to a magnificent and timeless inheritance, unbounded by our mortality and extending across the light years.

Those who are wise shall inherit the universe.

Bibliography

Ashpole, E., *Where is Everybody?* Sigma Press (1997).

Asimov, I., *Asimov's New Guide to Science*, Cambridge University Press (1992).

Barrow, J.D., *The World Within the World*, Oxford University Press (1988).

Barrow, J.D. and Tipler, F.J., *The Cosmological Anthropic Principle*, Oxford University Press (1996).

Bengston, S. (ed.), *Early Life on Earth*, Columbia University Press (1994).

Clark, S., *Towards the Edge of the Universe*, John Wiley (1997).

Cottingham, J., *Descartes*, Phoenix (1997).

Crosswell, K., *The Alchemy of the Heavens*, Oxford University Press (1995).

Deamer, D. and Fleischer, G.F., *Origins of Life: The Central Concepts,* Jones & Bartlett (1994).

Duve, C. de, *Vital Dust,* Basic Books (1995).

Drake, S., *Galileo Galilei*, Oxford University Press (1980).

English, N., 'The Cloudy World of Ancient Mars', *Astronomy Now*, Vol. 12, pp. 19-20 (1998).

English N., 'The Secret Lives of Comets', *Modern Astronomer*, Vol. 2, pp. 245-47 (1998).

Ferris, T., *Coming of Age in the Milky Way*, Vintage (1988).

Finch, C.E. and Tanzi, R.E., 'The Genetics of Aging', *Science*, 278, pp. 407-11 (1997).

Gjertsen, D., *Science and Philosophy*, Penguin (1989).

Hall, R., *Isaac Newton*, Cambridge University Press (1992).

Heidmann, J., *Extraterrestrial Intelligence*, Cambridge University Press (1992).

Honderich, T. (ed.), *The Oxford Companion to Philosophy*, Oxford University Press (1995).

Jones, J. (ed.), *Human Evolution*, Cambridge University Press (1992).

Kiernan, V. *et al.*, 'Did Martians Land in Antarctica?' *New Scientist*, Vol. 151, pp. 4-7, August (1996).

Kauffman, S., *At Home in the Universe*, Penguin (1995).

Kaufmann, W.J., *Universe*, W.H. Freeman and Company (1994).

Keeton, W.T., and Gould, J.L., *Biological Science*, W.W. Norton (1993).

Krings, M., *et al.*, 'Neanderthal DNA Sequences and the Origin of Modern Humans', *Cell*, Vol. 90, pp.19-30 (1997).

Lamberts, S.W., 'The Endocrinology of Aging', *Science*, 278, pp. 419-24 (1997).

Leakey, R. and Lewin, R., *Origins Reconsidered*, Abacus (1992).

Leakey, R. and Lewin, R., *The Sixth Extinction*, Phoenix Books (1995).

MacDougall, J.D., *A Short History of Planet Earth*, Wiley (1996).

McKay, S.D. *et al.*, 'Evidence for Past Life on Ancient Mars', *Science*, Vol. 273, pp. 924-30 (1996).

Mason, S.F., *Chemical Evolution*, Oxford University Press (1991).

Mills, R., *Space, Time and Quanta*, W.H. Freeman (1994).

Naeye, R., 'Was There Life on Mars?' *Astronomy*, Vol. 12, No. 11, pp. 46-53 (1996).

North, J., *The Fontana History of Astronomy and Cosmology*, Fontana (1994).

Sagan, C., *Cosmos*, Warner Books (1981).

Sagan, C., *Pale Blue Dot*, Headline (1994).

Sagan, C. and Shklovskii, I. S., *Intelligent Life in the Universe*, Holden Day (1966).

Silver, B.L., *The Ascent of Science*, Oxford University Press (1998).

Smolin, L., *The Life of the Cosmos*, Weidenfield & Nicolson (1997).

Stringer, C. and Gamble, C., *In Search of the Neanderthals*, James & Hudson (1993).

Stryer, L., *Biochemistry*, W.H. Freeman (1996).

Vaupel, J.W. *et al.*, 'Biodemographic Trajectories of Longevity', *Science*, 280, pp. 855-60 (1998).

Wilkinson, D., *Alone in the Universe?* Monarch Publications (1997).

Wolfson, R. and Pasachoff, J., *Physics*, HarperCollins (1997).

Young, L.B., *The Unfinished Universe*, Oxford University Press (1986).

Index